Pest Management in Vegetables

Pest Management in Vegetables

Part - II

Dr. I.M. Khan

AGROTECH PRESS

JAIPUR ▪ NEW DELHI

AGROTECH PRESS

Regd. Off. "Bony Residency", Gate No. 2
Opp. Tilak Public School, Vishweswariya Nagar,
Gopalpura Road, Jaipur - 302018, India
Ph: 0141 - 2761280, Telefax: 2761381
E-mail : agrotechpress@gmail.com

Delhi Office: 102, 1st Floor, Satyam House,
4327/3, Ansari Road, Darya Ganj,
New Delhi - 110002, India
Ph: 011-45652440
visit us: www.agrotechpress.com

Distributed By
KEYA INTERNATIONAL PRESS LLP
E-mail: keyainternationalpressllp@gmail.com

Published in India, 2024

ISBN : 9789384568436 (2 Vols. Set)

Typeset by Thomson Press

Printed and bound in India by Thomson Press Ltd.,
New Delhi

Contents

Preface

Vegetables are important components of daily diet throughout the world. Vegetables provide nutritional security in addition to food security. Their production extends to billions of tons and source of income for a large population. Yield and quality are central to sustainable vegetable production. If not properly managed, pests and diseases can dramatically reduce crop yield, quality and subsequent returns. A lot of time, money and natural resources are invested into growing vegetables. Even in the best-managed vegetable gardens - ones with soil rich in compost, and a diversity of plants to encourage natural predators - certain pests will occasionally get out of hand. Good pest and disease management can protect this investment from avoidable losses.

Limited tools to combat these problems is one of the major contributing factors to the high risk associated with vegetable production in general. Consequently, high level of management is required to successfully control pests' problems. Successful production of organically grown vegetables requires an even higher level of management for profitability to occur because even fewer control tools are available to producers using this production system. Organic producers have to rely heavily upon them minor details of vegetable culture in order to avoid problems. Pest control is often a result of the cumulative effects of many production practices. Emphasis is placed on prevention rather than control. This often is difficult to achieve with current technology. Therefore, a grower must develop pest control strategies. To do so a grower must have an understanding of the pest and the beneficials that usually occur in a given crop, and, he should also have the ability to anticipate pest problems and manipulate practices that minimize pests and maximize beneficial. Most of these tactics are used as preventive measures in an attempt to help keep pests below critical levels. Even after preventive measures are taken, certain conditions can still allow a pest population to build up to injurious levels. If pests do build up, then a rescue treatment can be used to reduce the population of the pest and rescue the crop. At the heart of pest management is the understanding that most crops can tolerate a certain amount of pest damage, thus a rescue treatment is not recommended until the pest population reaches a critical level that is referred to as a threshold.

Knowledge of which specific pests are present in the garden, along with background information about each pest's habits and life cycle, is the gardener's key to planning the most appropriate management strategy. Most insects found in gardens are not pests. Many are beneficial, preying on pests or performing other useful tasks. Correctly identifying the insects and other pests that attack vegetables is the first step toward controlling these pests effectively. One of the most important strategies for dealing with insects is to learn about insect life cycles, behaviors, habitats, and diets, and to recognize which are pests and which are actually lending you a helping hand. The pest descriptions in this book are designed to help the commercial vegetable grower and home gardener recognize the insects found on vegetables. Information on insect life cycles and habits is included to make pest control recommendations more understandable and usable.

Today, pests and diseases are managed using a more integrated approach. These approaches bring together the best mix of chemical and biological controls and cultural practices, to manage pests and diseases. They don't discard traditional chemical treatments and local knowledge, but integrate them into a sustainable system. Cultural practices are among the oldest techniques used for pest suppression, and many of the preventative practices used in conventional and organic farming today have their roots in traditional agriculture. Because only a limited range of suppressive pest control tactics are available to organic growers, knowledge-intensive cultural practices form the basis of organic pest management programs. Cultural, mechanical, physical, genetic, and biological controls help prevent severe pest problems, while pesticides are used when additional control measures are required. In a least-toxic pest control strategy a good step is to employ barriers, such as row covers, to exclude pests altogether. Using a pesticide, any pesticide, is always the measure of last resort. Gardeners today have at their disposal a handful of effective and safe pesticides.

Written with expertise from entomology, plant pathology, plant breeding, plant physiology, biochemistry, and extension education covering externalities of pesticide use, pest outbreaks, threshold theory, host plant resistance, crop plant manipulation, biological control, behavioral-modifying techniques, botanicals, and non-pesticide pest management, the book is an excellent source of advanced study material for academics, researchers and students, elucidating the underlying concepts of the pest management development process.

Dr. I.M. Khan

8

Jack-Fruit, Papaya and Nematode Problems

The jackfruit (*Artocarpus heterophyllus*, also known as jack tree, jakfruit, or sometimes simply jack or jak) is a species of tree in the *Artocarpus* genus of the mulberry family (*Moraceae*). It is native to parts of South and Southeast Asia, and is believed to have originated in the southwestern rain forests of India, in present-day Kerala, coastal Karnataka and Maharashtra. The jackfruit tree is well suited to tropical lowlands, and its fruit is the largest tree-borne fruit, reaching as much as 80 pounds (36 kg) in weight, 36 inches (90 cm) in length, and 20 inches (50 cm) in diameter.

The jackfruit tree is a widely cultivated and popular food item in tropical regions of India, Bangladesh, Nepal, Sri Lanka, Cambodia, Vietnam, Thailand, Malaysia, Indonesia, and the Philippines. Jackfruit is also found across Africa (e.g., in Cameroon, Uganda, Tanzania, Madagascar, and Mauritius), as well as throughout Brazil and in Caribbean nations such as Jamaica. Jackfruit is the national fruit of Bangladesh.

The word "jackfruit" comes from Portuguese *jaca*, which in turn, is derived from the Malayalam language term, *chakka* (Malayalam Chakka pazham : M). When the Portuguese arrived in India at Kozhikode (Calicut) on the Malabar Coast (Kerala) in 1498, the Malayalam name *chakka* was recorded by Hendrik van Rheede (1678–1703) in the *Hortus Malabaricus*, vol. iii in Latin. Henry Yule translated the book in Jordanus Catalani's (f. 1321–1330) *Mirabilia descripta: the wonders of the East.*

The common English name "jackfruit" was used by the physician and naturalist Garcia de Orta in his 1563 book *Colóquios dos simples e drogas da India.* Centuries later, botanist Ralph Randles Stewart suggested it was named after William Jack (1795–1822), a Scottish botanist who worked for the East India Company in Bengal, Sumatra, and Malaysia. This could not be true, as the fruit was called a "jack" in English before William Jack was born: for instance, in Dampier's 1699 book, *A New Voyage Round the World.* It is called Kathal in Bengali, Katahal in Hindi, Pala-pazham in Tamil, Panasa in Telugu, Phanas in Marathi and Gujarati, Halasu in Karnataka, Fenesi in Kiswahili, Nangka in Malay and Indonesian, Langka in the Philippines, and Ka-noon in Thailand.

Synonym Discussion

Artocarpus integer *(Thunb.) Merr. is currently accepted name, whereas* Artocarpus integrifolius *L.f. is synonym. However in Flora of British India, Volume 5 (Page 541), J.D. Hooker mentions it as* Artocarpus integrifolia *L.f. Moreover,* Artocarpus heterophyllus *Lam. is a different species.*

Cultivation and Ecology

Figure: *Jackfruit garden in Bangladesh*

The jackfruit has played a significant role in Indian agriculture for centuries. Archeological findings in India have revealed that jackfruit was cultivated in India 3000 to 6000 years ago. It is also widely cultivated in southeast Asia.

Thailand and Vietnam are major producers of jackfruit, a lot of which are cut, prepared and canned in a sugary syrup (or frozen in bags/

boxes without syrup), and exported overseas, frequently to North America and Europe.

In other areas, the jackfruit is considered an invasive species as in Brazil's Tijuca Forest National Park in Rio de Janeiro. The Tijuca is mostly an artificial secondary forest, whose planting began during the mid-19th century, and jackfruit trees have been a part of the park's flora since its founding. Recently, the species has expanded excessively; its fruits, which naturally fall to the ground and open, are eagerly eaten by small mammals such as the common marmoset and coati.

The seeds are dispersed by these animals, which allows the jackfruit to compete for space with native tree species. Additionally, as the marmoset and coati also prey opportunistically on bird's eggs and nestlings, the supply of jackfruit as a ready source of food has allowed them to expand their populations, to the detriment of the local bird populations. Between 2002 and 2007, 55,662 jackfruit saplings were destroyed in the Tijuca Forest area in a deliberate culling effort by the park's management.

Soil and Climate

Although the jackfruit is essentially a tree of the tropical lowlands, it is adapted to a wider range of conditions. It can tolerate higher altitudes and cold better than the breadfruit.

The jackfruit can be grown on a variety of soils as long as they are well-drained, but does best in deep alluvial soils of open texture. Most of the trees grown in Trengganu and Pahang are on the Holyrood series.

Propagation

The most common and simplest method of raising jackfruit trees is from seeds, though other methods are also known in Malaysia. The trees do not generally breed true from seeds, so the practice has led to immense variation in yields, fruit characters and quality. Inarching and grafting by the Forkert Method is known in Malaysia as a means of perpetuating desirable clonal stocks.

Planting

The place chosen for the planting should be first cleared from old tree stumps and old roots to avoid termites and root disease. When necessary, the soil should be ploughed first, then rows are made to mark the planting intervals. Usually, jackfruits are planted at a distance of 30 ft. x 30 ft. In an acre, 48 trees can be planted. In a new area the planting interval can be reduced to 25 ft. x 25 ft., and 69 trees can be planted in an acre.

Planting holes of 2 ft x 2 ft x 2 ft., should be prepared, and top soil with 4 oz. CIRP should be added into each hole. For not every fertile soils, it is advisable to add 40 lbs. of cowdung to the top soil. Usually the planting holes are left open for fourteen (14) days before they are filled up again, and only then is the budgrafted jackfruit tree planted. It is important to remember that during planting, the bud patch is not to be covered with soil.

It would otherwise cause the bud patch to rot and die. The amount of sunlight can be reduced by using shade from coconut fronds. Usually, bud-grafted trees are planted during the rainy season so that they do not have to be watered. Since watering is quite a problem in large scale cultivation, planting should be done when there is rain in order to make sure that the plants can grow well. This is important to prevent the plants from being stunted. Shading from the coconut fronds can be removed after two weeks if the weather is fine. Otherwise, it should be left for another week or more.

For sandy soil and clay soil or the Holyrood series, legumes are required as cover crops. Calopogonium, Centrosema and Pueraria in a ratio of 5:4:1 are usually used. Cover crops are required to prevent weeds from growing, to alter the condition and fertility of the soil and also to prevent the soil from becoming too hot especially in bris areas.

Fertilizer and Manuring

A consistent, well-balanced manurial programme is important so as to stimulate rapid growth in young trees and to ensure maximum yield when the plants come into bearing. As nitrogen, phosphate and potash play a vital role in the plant metabolism, and markedly affect fruit production, a balanced supply of these nutrients in the fertilizer mixture must be applied to the plant.

The amount and time of fertilizer application are as shown below:

Year	*Fertilizer mixture*	*Amount per tree/ application*	*Time of application*
1st year	G	4 oz	every 3 months
2nd year	G	6 oz	every 3 months
3rd year	F	1 lb.	every 4 months
4th year	F	2 lb.	every 4 months
5th year	F	3 lb.	every 6 months
6th year	F	5 lb.	every 6 months
7th year onwards	F	6-7 lb.	every 6 months

G = Growth mixture

F = Flowering mixture

Aroma

Jackfruit are known for having a distinct aroma. In a study using five jackfruit cultivars, the main jackfruit volatile compounds that were detected are: ethyl isovalerate, 3-methylbutyl acetate, 1-butanol, propyl isovalerate, isobutyl isovalerate, 2-methylbutanol, and butyl isovalerate. These compounds were consistently present in all the five cultivars studied, suggesting that these esters and alcohols contributed to the sweet and fruity aroma of jackfruit.

Fruit

Figure: *Jackfruit Flesh*

The flesh of the jackfruit is starchy and fibrous and is a source of dietary fibre. The flavour is comparable to a combination of apple, pineapple, mango and banana. Varieties are distinguished according to characteristics of the fruit's flesh. In Brazil, three varieties are recognised: *jaca-dura*, or the "hard" variety, which has a firm flesh and the largest fruits that can weigh between 15 and 40 kilograms each, *jaca-mole*, or the "soft" variety, which bears smaller fruits with a softer and sweeter flesh, and *jaca-manteiga*, or the "butter" variety, which bears sweet fruits whose flesh has a consistency intermediate between the "hard" and "soft" varieties. In Indochina, there are 2 varieties, being the "hard" version (more crunchy, drier and less sweet but fleshier), and the "soft" version (more soft, moister, much sweeter with a darker gold-colour flesh than the hard variety).

In Kerala, two varieties of jackfruit predominate: *varikka* and *koozha* . *Varikka* has a slightly hard inner flesh when ripe, while the inner flesh of the ripe *koozha* fruit is very soft and almost dissolving. A sweet preparation called *chakka varattiyathu* (jackfruit jam) is made by seasoning pieces of *varikka* fruit flesh in jaggery, which can be preserved and used for many months. Huge jackfruits up to four feet in length with a corresponding girth are sometimes seen in Kerala.

In West Bengal the two varieties are called *khaja kathal* and *moja kathal*. The fruits are either eaten alone or as a side to rice / roti / chira / muri. Sometimes the juice is extracted and either drunk straight or as a side with muri. The extract is sometimes condensed into rubbery delectables and eaten as candies. The seeds are either boiled or roasted and eaten with salt and hot chillies. They are also used to make spicy side-dishes with rice or roti.

In Mangalore, Karnataka, the varieties are called *bakke* and *imba*. The pulp of the *imba* jackfruit is ground and made into a paste, then spread over a mat and allowed to dry in the sun to create a natural chewy candy.

The young fruit is called *polos* in Sri Lanka and *idichakka* or *idianchakka* in Kerala.

In Indochina, jackfruit is a frequent ingredient in sweets and desserts. In Vietnam, jackfruit is used to make jackfruit Chè (*chè* is a sweet dessert soup, similar to the Chinese derivative, *bubur chacha*). The Vietnamese also use jackfruit puree as part of pastry fillings, or as a topping on Xôi ngÍt (sweet version of sticky rice portions).

Ripe jackfruit is naturally sweet with subtle flavouring. It can be used to make a variety of dishes, including custards, cakes, *halo-halo* and more. In India, when the Jackfruit is in season, an ice cream chain store called "Naturals" carries Jackfruit flavoured ice cream.

Ripe jackfruit arils are sometimes seeded, fried or freeze-dried and sold as jackfruit chips.

The seeds from ripe fruits are edible, are said to have a milky, sweet taste, and may be boiled, baked or roasted. When roasted the flavour of the seeds is comparable to chestnuts. Seeds are used as snacks either by boiling or fire roasted, also used to make desserts. For making the traditional breakfast dish in southern India: idlis, the fruit is used along with rice as an ingredient and jackfruit leaves are used as a wrapping for steaming. Jackfruit dosas can be prepared by grinding jackfruit flesh along with the batter.

Culinary Uses for Unripe Fruit

The cuisines of India, Nepal, Bangladesh, Sri Lanka, Indonesia, Cambodia, Thailand and Vietnam use cooked young jackfruit. In Indonesia, young jackfruit is cooked with coconut milk as *gudeg*. In many cultures, jackfruit is boiled and used in curries as a staple food. In northern Thailand, the boiled young jackfruit is used in the Thai salad called *tam kanun*. In West Bengal the unripe green jackfruit called "aechor/ichor" is used as a vegetable to make various spicy curries, side-dishes and as fillings for cutlets & chops. It is especially sought after by vegetarians who substitute this for meat and hence is nicknamed as gacch-patha (tree-mutton). In the Philippines, it is cooked with coconut milk (*ginataang langka*). In Réunion Island, it is cooked either alone or with animal flesh, such as shrimp or smoked pork. In southern India unriped Jackfruit slices are deep fried to make chips. In Udipi cuisine Jack fruit is used make appa and Addae.

Figure: *Green jackfruit and potato curry, Kolkata.*

Because unripe jackfruit has a meat-like taste, it is used in curry dishes with spices, in Bihar, Jharkhand, Sri Lankan, Andhran, eastern-Indian (Bengali) and (Odisha) and Keralan cuisine. The skin of unripe jackfruit must be peeled first, then the remaining whole jackfruit can be chopped into edible portions and cooked before serving. Young jackfruit has a mild flavour and distinctive meat-like texture and is compared to poultry. Meatless sandwiches have been suggested and are popular with both vegetarian and nonvegetarian populations. Unripe jackfruit is widely known as *Panasa Katha* in Odisha.

Nutrition

The edible jackfruit is made of soft, easily-digestible flesh (bulbs); A portion of 100 g of edible raw jackfruit provides about 95 calories and is a good source of the antioxidant vitamin C, providing about 13.7 mg. Jackfruit seeds are rich in protein. The fruit is also rich in potassium, calcium, and iron.

Seeds

In general, the seeds are gathered from the ripe fruit, sun-dried, then stored for use in rainy season in many parts of South Indian states.

They are extracted from fully matured fruits and washed in water to remove the slimy part. Seeds should be stored immediately in closed polythene bags for one or two days to prevent them from drying out. Germination is improved by soaking seeds in clean water for 24 hours. During transplanting, sow seeds in line, 30 cm apart, in a nursery bed filled with 70% soil mixed with 30% organic matter. The seedbed should be shaded partially from direct sunlight in order to protect emerging seedlings.

Boiled Jackfruit seed is also edible. Seasoned with nothing more than salt, this snack is very popular in Java.

Wood

Figure: *Jackfruit tree*

The wood of the tree is used for the production of musical instruments. In Indonesia, hardwood from the trunk is carved out to form the barrels of drums used in the *gamelan*, and in the Philippines its soft wood is made into the body of the *kutiyapi*, a type of boat lute. It is also used to make the body of the Indian string instrument *veena* and the drums *mridangam*, *thimila* and *kanjira*; the golden, yellow timber with good grains is used for building furniture and house construction in India. The ornate wooden plank called *avani palaka* made of the wood of jackfruit tree is used as the priest's seat during Hindu ceremonies in Kerala. In Vietnam, jackfruit wood is prized for the making of Buddhist statuaries in temples.

Jackfruit wood is widely used in the manufacture of furniture, doors and windows, and in roof construction. The heartwood is used by Buddhist forest monastics in Southeast Asia as a dye, giving the robes of the monks in those traditions their distinctive light-brown colour.

Commercial Availability

Outside of its countries of origin, fresh jackfruit can be found at Asian food markets, especially in the Philippines, Thailand, Vietnam, Malaysia Cambodia and Bangladesh. It is also extensively cultivated in the Brazilian coastal region, where it is sold in local markets. It is available canned in sugary syrup, or frozen, already prepared and cut. Dried jackfruit chips are produced by various manufacturers. In northern Australia, particularly in Darwin, jackfruit can be found at outdoor produce markets during the dry season. Outside of countries where it is grown, jackfruit can be obtained year-round both canned or dried. It has a ripening season in Asia of late spring to late summer.

There are established jackfruit industries in Sri Lanka and Vietnam, where the fruit is processed into products such as flour, noodles, papad and ice cream. It is also canned and sold as a vegetable for export.

Pests and Diseases

Jackfruit is seldom attacked by serious fungal or virus diseases. The common diseases which occur are physiological, where the trees wilt and the shoots die. This is usually caused by shortage of water, flooded or waterlogged conditions. Another disease which can reduce the production of jackfruit is caused by fungal attack during the development of the fruits. This disease can be controlled by wrapping the fruits before they are attacked by fruit borers, by removing and burning the rotten flowers and fruits and also by spraying with fungicide such as Perenox, Vitigran blue twice a week during the early fruiting season.

Natural fruit drop has also been observed. Many fruit buds which are still small drop even though they are not attacked by insects or fungus. It is believed that this is due to some physiological reasons such as lack of nutrients. A few diseases caused by insects and pests such as jackfruit borer, fruit flies, civet cat and wild boar have been reported.

Jackfruit Borer

Margaronia *sp. is a type of Lepidopteran which can destroy every part of a jackfruit tree. They usually lay eggs on leaves and fruits. The larvae eat the leaves at the beginning and later bore into fruits and develop. After destroying part of the fruit they begin to bore into the branches and stem.*

The method of controlling Margaronia *larvae is to spray insecticides such as carbaryl and gamma BHC at least once a fortnight at a concentration of 0.1% of the active ingredient. For the plants which have already been attacked, it is necessary to collect and destroy the rotten fruits; the dead branches or stems have to be cut. The holes in the fresh part should be filled with paradichloro-benzene or other insecticides such as gamma BHC.*

Fruit Flies

Two types of fruit flies, Dacus umbrosus *and* Dacus dorsata *can affect the jackfruit production. Fruit flies breed in the matured fruit; the adult female lays eggs at the inner part of the skin and the larvae eat and thrive inside the fruit which later becomes rotten.*

The method of controlling fruit flies is by destroying the rotten fruits, either by burning or burying them in the soil. Covering fruits using coconut leaf bags, gunny sacks or papers is good and effective.

A new method of controlling fruit flies is to use poisoned attractants such as methyl eugenol and protein hydrolysate mixed with trichlorphon (Dipterex) but the method does not give satisfactory control.

Mammalian Pests

Jackfruit fruits are loved by wild animals such as the civet cat. Another pest which can cause destruction and loss is the wild boar. All immature and mature fruits at the lower part of the trees will be destroyed by these pests. Besides that, they like to dig the roots to find insects and this may cause the trees to fall. A method of controlling them is by elimination by shooting or by using poisonous bait such as 'methomyl' (Lannate) inside pineapple or sweet potatoes.

Papaya

The papaya (from Carib via Spanish), papaw, or pawpaw is the fruit of the plant Carica papaya, the sole species in the genus Carica of the plant family Caricaceae. It is native to the tropics of the Americas, perhaps from southern Mexico and neighbouring Central America. It was first cultivated in Mexico several centuries before the emergence of the Mesoamerican classical civilizations.

The papaya is a large, tree-like plant, with a single stem growing from 5 to 10 m (16 to 33 ft) tall, with spirally arranged leaves confined to the top of the trunk. The lower trunk is conspicuously scarred where leaves and fruit were borne. The leaves are large, 50–70 cm (20–28 in) in diameter, deeply palmately lobed, with seven lobes. Unusually for such large plants, the trees are dioecious. The tree is usually unbranched, unless lopped. The flowers are similar in shape to the flowers of the Plumeria, but are much smaller and wax-like. They appear on the axils of the leaves, maturing into large fruit - 15–45 cm (5.9–17.7 in) long and 10–30 cm (3.9–11.8 in) in diameter. The fruit is ripe when it feels soft (as soft as a ripe avocado or a bit softer) and its skin has attained an amber to orange hue.

Carica papaya *was the first transgenic fruit tree to have its genome deciphered.*

Common Names

Carica papaya *plants and their fruits are known by different names around the English-speaking world:*

- North America and Belize: papaya
- United Kingdom: The fruit is usually called papaya but is also known as papaw or pawpaw
- Africa: pawpaw or papaw

Distribution

Carica papaya is native to the New World in Belize, Bolivia, Brazil, Colombia, Costa Rica, Ecuador, El Salvador, French Guiana, Guatemala, Guyana, Honduras, Mexico, Nicaragua, Panama, Paraguay, Peru, Suriname, Venezuela, and northern Argentina. C. papaya has become naturalized in the Bahamas, Cuba, the Dominican Republic, Grenada, Guadeloupe, Haiti, Jamaica, Martinique, Puerto Rico, Saint Vincent and the Grenadines, Trinidad and Tobago, the U.S. Virgin Islands, the U.S. state of Florida, and Malawi and Tanzania in Africa. Additional crops of C. papaya are grown in Australia, the Philippines, and the U.S. state of Hawaii.

Cultivation

Gaining in popularity among tropical fruits worldwide, papaya is now ranked third with 11.22 Mt, or 15.36 percent of the total tropical fruit production, behind mango with 38.6 Mt (52.86%) and pineapple with 19.41 Mt (26.58%). Global papaya production has grown significantly over the last few years, mainly as a result of increased production in India. Papaya has become an important agricultural export for developing countries, where export revenues of the fruit provide a livelihood for thousands of people, especially in Asia and

Latin America. Papaya exports contribute to the growing supply of healthy food products on international markets. The top three exporting countries accounted for 63.28 percent of the total global exports of papaya between 2007 and 2009, with more than half of those exports going to the United States.

Global papaya production is highly concentrated, with the top ten countries averaging 86.32 percent of the total production for the period 2008–2010. India is the leading papaya producer, with a 38.61 percent share of the world production during 2008–2010, followed by Brazil (17.5%) and Indonesia (6.89%). Other important papaya producing countries and their share of global production include Nigeria (6.79%), Mexico (6.18%), Ethiopia (2.34%), Democratic Republic of the Congo (2.12%), Colombia (2.08%), Thailand (1.95%), and Guatemala (1.85%).

Originally from southern Mexico (particularly Chiapas and Veracruz), Central America, and northern South America, the papaya is now cultivated in most tropical countries. In cultivation, it grows rapidly, fruiting within three years. It is, however, highly frost-sensitive, limiting its production to tropical climates. Temperatures below "2 °C (29 °F) are greatly harmful if not fatal. In Florida, growth is generally limited to southern parts of the state. It also prefers sandy, well-drained soil as standing water will kill the plant within 24 hours.

For cultivation, however, only female plants are used, since they give off a single flower each time, and close to the base of the plant, while the male gives off multiple flowers in long stems, which result in poorer quality fruit.

Cultivars

Two kinds of papayas are commonly grown. One has sweet, red or orange flesh, and the other has yellow flesh; in Australia, these are called "red papaya" and "yellow papaw", respectively. Either kind, picked green, is called a "green papaya."

The large-fruited, red-fleshed 'Maradol', 'Sunrise', and 'Caribbean Red' papayas often sold in US markets are commonly grown in Mexico and Belize.

In 2011 Philippine researchers reported that by hybridizing papaya with *Vasconcellea quercifolia* they had developed conventionally bred, nongenetically engineered papaya resistant to PRV.

Genetically Engineered Cultivars

In response to the papaya ringspot virus (PRV) outbreak in Hawaii, genetically altered papaya were generated and brought to market (including 'SunUp' and 'Rainbow') that have some PRV DNA incorporated into the DNA of the plant are resistant to PRVs. This was so successful that by 2010, 80% of Hawaiian papaya plants were genetically modified.

Uses

Papayas can be used as a food, a cooking aid and in traditional medicine. The stem and bark may be used in rope production.

Meat Tenderizing

Both green papaya fruit and the tree's latex are rich in papain, a protease used for tenderizing meat and other proteins. Its ability to break down tough meat fibres was used for thousands of years by indigenous Americans. It is now included as a component in powdered meat tenderizers.

Nutrients, Phytochemicals and Culinary Practices

Papaya fruit is a source of nutrients such as provitamin A carotenoids, vitamin C, folate and dietary fibre. Papaya skin, pulp and seeds also contain a variety of phytochemicals, including lycopene and polyphenols. In preliminary research, danielone, a phytoalexin found in papaya fruit, showed antifungal activity against *Colletotrichum gloesporioides*, a pathogenic fungus of papaya.

The ripe fruit of the papaya is usually eaten raw, without skin or seeds. The unripe green fruit can be eaten cooked, usually in curries, salads, and stews. Green papaya is used in Southeast Asian cooking, both raw and cooked. In Thai cuisine, papaya is used to make Thai salads such as *som tam* and Thai curries such as *kaeng som* when still not fully ripe. In Indonesian cuisine, the unripe green fruits and young leaves are boiled for use as part of *lalab* salad, while the flower buds are sautéed and stir-fried with chillies and green tomatoes as Minahasan papaya flower vegetable dish. Papayas have a relatively high amount of pectin, which can be used to make jellies. The smell of ripe, fresh papaya flesh can strike some people as unpleasant. In Brazil, the unripe fruits are often used to make sweets or preserves.

The black seeds of the papaya are edible and have a sharp, spicy taste. They are sometimes ground and used as a substitute for black pepper. In some parts of Asia, the young leaves of the papaya are steamed and eaten like spinach.

Traditional Medicine

In some parts of the world, papaya leaves are made into tea as a treatment for malaria. Antimalarial and antiplasmodial activity has been noted in some preparations of the plant, but the mechanism is not understood and no treatment method based on these results has been scientifically proven. In belief that it can raise platelet levels in blood, papaya may be used as a medicine for dengue fever. Papaya is marketed in tablet form to remedy digestive problems.

Papain is also applied topically for the treatment of cuts, rashes, stings and burns. Papain ointment is commonly made from fermented papaya flesh, and is applied as a gel-like paste. Harrison Ford was treated for a ruptured disc incurred during filming of *Indiana Jones and the Temple of Doom* by papain injections.

Women in India, Bangladesh, Pakistan, Sri Lanka, and other countries have long used green papaya as an herbal medicine for contraception and abortion. Enslaved women in the West Indies were noted for consuming papaya to prevent pregnancies and thus preventing their children from being born into slavery.

Preliminary Research

Preliminary research in animals has provided evidence for the potential contraceptive and abortifacient capability of papaya. Laboratory studies have also shown that papaya seeds have contraceptive effects in adult male langur monkeys, and possibly in adult male humans. Unripe papaya is especially effective in large amounts or high doses. Ripe papaya is not teratogenic and will not cause miscarriage in small amounts. Phytochemicals in papaya may suppress the effects of progesterone.

Other preliminary research indicates alternate possible effects which remain to be further studied. Papaya juice has an *in vitro* antiproliferative effect on liver cancer cells, possibly due to lycopene or immune system stimulation. Papaya seeds might contain antibacterial properties against *Escherichia coli*, *Staphylococcus aureus* or *Salmonella typhi*. Papaya seed extract may have effects in toxicity-induced kidney failure.

Allergies and Side Effects

Papaya is frequently used as a hair conditioner, but should be used in small amounts. Papaya releases a latex fluid when not quite ripe, which can cause irritation and provoke allergic reaction in some people.

The latex concentration of unripe papayas is speculated to cause uterine contractions, which may lead to a miscarriage. Papaya seed extracts in large doses have a contraceptive effect on rats and monkeys, but in small doses have no effect on the unborn animals.

Excessive consumption of papaya can cause carotenemia, the yellowing of soles and palms, which is otherwise harmless. However, a very large dose would need to be consumed; papaya contains about 6% of the level of beta carotene found in carrots (the most common cause of carotenemia).

Diseases and Pests

Viruses

Papaya ringspot virus is a well-known virus within plants in Florida. The first signs of the virus are yellowing and vein-clearing of younger leaves as well as mottling yellow leaves. Infected leaves may obtain blisters, roughen or narrow, with blades sticking upwards from the middle of the leaves. The petioles and stems may develop dark green greasy streaks and in time become shorter. The ringspots are circular, C-shaped markings that are darker green than the fruit itself. In the later stages of the virus, the markings may become gray and crusty. Viral infections impact growth and reduce the fruit's quality. One of the biggest effects that viral infections have on papaya is the taste. Right now, the virus is uncontrollable.

The papaya mosaic virus is very destructive because it completely destroys the plant until only a small tuft of leaves are left. The virus affects both the leaves of the plant and also the fruit itself. Leaves show thin, irregular, dark-green lines around the borders and clear areas around the veins. The more severely affected leaves are irregular and linear in shape. The virus can infect the fruit at any stage of its maturity. Fruits as young as 2 weeks old have been spotted with dark-green ringspots about 1 inch in diameter. Rings on the fruit are most likely seen on either the stem end or the blossom end. In the early stages of the ringspots, the rings tend to be many closed circles, but as the disease develops the rings will increase in diameter consisting of one large ring. The difference between the ringspot and the mosaic viruses is the ripe fruit in the ringspot has mottling of colours and mosaic doesn't.

Fungi

The fungus Anthracnose is known to specifically attack papaya especially the mature fruits. The disease starts out small with very few signs, such as water-soaked spots on ripening fruits. The spots become sunken, turn brown or black and may get bigger. In some of the older spots, the fungus may produce pink spores. The fruit ends up being soft and having an off flavour because the fungus grows into the fruit.

The fungus powdery mildew occurs as a superficial white presence on the surface of the leaf in which it is easily recognised. Tiny, light yellow spots begin on the lower surfaces of the leaf as the disease starts to make its way. The spots enlarge and white powdery growth appears on the leaves. The infection usually appears at the upper leaf surface as white fungal growth. Powdery Mildew is not as severe as other diseases.

The fungus phythphthora blight causes damping-off, root rot, stem rot, stem girdling and fruit rot. Damping-off happens in very young plants by wilting and death in plant. The spots on established plants start out as water-soaked lesions at the fruit and branch scars. These spots can get bigger and cause the death of the plant. The roots can be severely and rapidly infected, causing the plant to rapidly brown and wilt away collapsing within days. The most dangerous feature of the disease is the infection of the fruit because it cause harm to people who consume it. The biggest evidence that the fungus is present are the water-soaked marks that appears first along with the white fungus that grows on the dead fruit. After the fruit dies it shrivels and falls to the ground.

Pests

The papaya fruit fly is mainly yellow with black marks. The female papaya fruit fly has a very long, slender abdomen with an extended ovipositor that exceeds the length of its body. The male papaya fruit fly looks like the female with the differences of a hairy abdomen and no ovipositor. Long slender eggs are laid inside of the fruit by the female papaya fruit fly. The larva are white and look very much like the regular fruit fly larvae. The female is capable of laying up to 100 or more eggs and are laid during the evening or early morning in groups of ten inside young fruit. They usually hatch within 12 days of being in the fruit where they'll feed on the seeds and interior parts of the fruit. When the larvae matures (usually 16 days after being hatched) they eat their way out of the fruit, drop to the ground, and pupate just below the soil and emerge within one to two weeks as mature flies. The flesh of the papaya must be ripe in order for the fly to migrate towards the surface of the fruit because unripe papaya juice is fatal to them. The papaya will turn yellow and drop to the ground if it is infected by the papaya fruit fly.

The two-spotted spider mite is a 0.5 mm long brown or orange-red but a green, greenish yellow translucent oval pest. They all have needle-like piercing-sucking mouthparts and feed by piercing the plant tissue with their mouth parts usually the underside of the plant. The spider mites spin fine threads of webbing on the host plant and when they remove the sap, the mesophyll tissue collapses and a small chlorotic spot forms at the feeding sites. The leaves of the papaya fruit turn yellow, gray or bronze. If the spider mites aren't controlled it can cause the death of the fruit.

The papaya whitefly lays yellow oval eggs that appear dusted on the undersides of the leaves. They eat the papaya fruits leaves therefore damaging the fruit. There, the eggs developed into flies in three stages called instars. The first instar has well-developed legs and is the only mobile immature life stage. The crawlers insert their mouthparts in the lower surfaces of the leaf when they find it suitable and usually don't move again in this stage. The next instars are flattened, oval and scale-like. In the final stage as the pupal the whiteflies are more convex, with large conspicuous red eyes.

Nematodes

The nematodes or roundworms constitute the phylum Nematoda. They are a diverse animal phylum inhabiting a very broad range of environments. Nematode species can be difficult to distinguish, and although over 25,000 have been described, of which more than half are parasitic, the total number of nematode species has been estimated to be about 1 million. Unlike cnidarians and flatworms, nematodes have tubular digestive systems with openings at both ends.

Habitats

Nematodes have successfully adapted to nearly every ecosystem from marine to fresh water, to soils, and from the polar regions to the tropics, as well as the highest to the lowest of elevations. They are ubiquitous in freshwater, marine, and terrestrial environments, where they often outnumber other animals in both individual and species counts, and are found in locations as diverse as mountains, deserts and oceanic trenches. They are found in every part of the earth's lithosphere. They represent, for example, 90% of all life forms on the ocean floor. Their numerical dominance, often exceeding a million individuals per square metre and accounting for about 80% of all individual animals on earth, their diversity of life cycles, and their presence at various trophic levels point at an important role in many ecosystems. Their many parasitic forms include pathogens in most plants and animals (including humans). Some nematodes can undergo cryptobiosis.

One group of carnivorous fungi, the nematophagous fungi, are predators of soil nematodes. They set enticements for the nematodes in the form of lassos or adhesive structures. Nematodes have even been found at great depth (0.9–3.6 km) below the surface of the Earth in gold mines in South Africa.

Nathan Cobb described the ubiquity of nematodes on Earth thus:

In short, if all the matter in the universe except the nematodes were swept away, our world would still be dimly recognisable, and if, as disembodied spirits, we could then investigate it, we should find its mountains, hills, vales, rivers, lakes, and oceans represented by a film of nematodes. The location of towns would be decipherable, since for every massing of human beings there would be a corresponding massing of certain nematodes. Trees would still stand in ghostly rows representing our streets and highways. The location of the various plants and animals would still be decipherable, and, had we sufficient knowledge, in many cases even their species could be determined by an examination of their erstwhile nematode parasites."

Taxonomy and Systematics

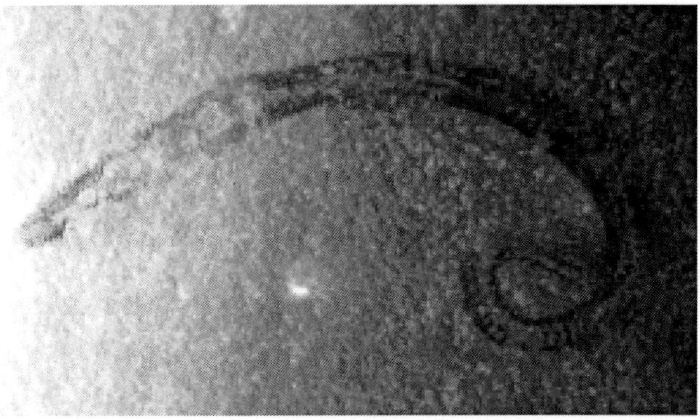

Figure: Eophasma jurasicum, *a fossilized nematode*

The group was originally defined by Karl Rudolphi in 1808 under the name Nematoidea, from Ancient Greek íÆìá (*nêma, nêmatos*, 'thread') and -eiä ò (*-eidçs*, 'species'). It was reclassified as family Nematodes by Burmeister in 1837 and order Nematoda by K. M. Diesing in 1861.

At its origin, the "Nematoidea" included both roundworms and horsehair worms. Along with Acanthocephala, Trematoda and Cestoidea, it formed the group Entozoa. The first differentiation of roundworms from horsehair worms, though erroneous, is due to von Siebold (1843) with orders Nematoidea and Gordiacei (Gordiacea). They were classed along with Acanthocephala in the new phylum Nemathelminthes (today obsolete) by Gegenbaur (1859). The taxon Nematoidea, including the family Gordiidae (horsehair worms), was then promoted to the rank of phylum by Ray Lankester (1877). In 1919, Nathan Cobb proposed that roundworms should be recognised alone as a phylum. He argued they should be called nema(s) in English rather than "nematodes" and defined the taxon Nemates (Latin plural of *nema*). Since Cobb was the first to exclude all but nematodes from the group, some sources consider the valid taxon name to be Nemates or Nemata, rather than Nematoda.

Phylogeny

The relationships of the nematodes and their close relatives among the protostomian Metazoa are unresolved. Traditionally, they were held to be a lineage of their own but in the 1990s, they were proposed to form the group Ecdysozoa together with moulting animals, such as arthropods. The identity of the closest living relatives of the Nematoda has always been considered to be well resolved. Morphological characters and molecular phylogenies agree with placement of the roundworms as a sister taxon to the parasitic horsehair worms (Nematomorpha); together they make up the Nematoida. Together with the Scalidophora (formerly Cephalorhyncha), the Nematoida form the Introverta. It is entirely unclear whether the Introverta are, in turn, the closest living relatives of the enigmatic Gastrotricha; if so, they are considered a clade Cycloneuralia, but much disagreement occurs both between and among the available morphological and molecular data. The Cycloneuralia or the Introverta—depending on the validity of the former—are often ranked as a superphylum.

Nematode Systematics

Due to the lack of knowledge regarding many nematodes, their systematics is contentious. An earliest and influential classification was proposed by Chitwood and Chitwood—later revised by Chitwood—who divided the phylum into two—the Aphasmidia and the Phasmidia. These were later renamed Adenophorea (gland bearers) and Secernentea (secretors), respectively. The Secernentea share several characteristics, including the presence of phasmids, a pair of sensory organs located in the lateral posterior region, and this was used as the basis for this division. This scheme was adhered to in many later classifications, though the Adenophorea were not a uniform group.

Initial DNA sequence studies suggested the existence of five clades:

- Dorylaimia
- Enoplia
- Spirurina
- Tylenchina
- Rhabditina

As it seems, the Secernentea are indeed a natural group of closest relatives. But the "Adenophorea" appear to be a paraphyletic assemblage of roundworms simply retaining a good number of ancestral traits. The old Enoplia do not seem to be monophyletic either, but to contain two distinct lineages.

The old group "Chromadoria" seem to be another paraphyletic assemblage, with the Monhysterida representing a very ancient minor group of nematodes. Among the Secernentea, the Diplogasteria may need to be united with the Rhabditia, while the Tylenchia might be paraphyletic with the Rhabditia.

The understanding of roundworm systematics and phylogeny as of 2002 is summarised below:

Phylum Nematoda

- Basal order Monhysterida
- Class Dorylaimea
- Class Enoplea
- Class Secernentea
    - Subclass Diplogasteria (disputed)
    - Subclass Rhabditia (paraphyletic?)
    - Subclass Spiruria
    - Subclass Tylenchia (disputed)
- "Chromadorea" assemblage

Later work has suggested the presence of 12 clades. The Secernentea—a group that includes virtually all major animal and plant 'nematode' parasites—apparently arose from within the Adenophorea.

A major effort to improve the systematics of this phylum is in progress and being organised by the 959 Nematode Genomes.

A complete checklist of the World's nematode species can be found in the World Species Index:Nematoda.

An analysis of the mitochondrial DNA suggests that the following groupings are valid

- subclass Dorylaimia
- orders Rhabditida, Trichinellida and Mermithida
- suborder Rhabditina
- infraorders Spiruromorpha and Oxyuridomorpha

The Ascaridomorpha, Rhabditomorpha and Diplogasteromorpha appear to be related.

The suborders Spirurina and Tylenchina and the infraorders Rhabditomorpha, Panagrolaimomorpha and Tylenchomorpha are paraphytic.

The monophyly of the Ascaridomorph is uncertain.

Anatomy

Nematodes are slender worms: typically approximately 5 to 100 μm thick, and at least 0.1 mm (0.0039 in) but less than 2.5mm long. The smallest nematodes are microscopic, while free-living species can reach as much as 5 cm (2.0 in), and some parasitic species are larger still, reaching over a metre in length. The body is often ornamented with ridges, rings, bristles, or other distinctive structures.

The head of a nematode is relatively distinct. Whereas the rest of the body is bilaterally symmetrical, the head is radially symmetrical, with sensory bristles and, in many cases, solid 'head-shields' radiating outwards around the mouth. The mouth has either three or six lips, which often bear a series of teeth on their inner edges. An adhesive 'caudal gland' is often found at the tip of the tail.

The epidermis is either a syncytium or a single layer of cells, and is covered by a thick collagenous cuticle. The cuticle is often of complex structure, and may have two or three distinct layers. Underneath the epidermis lies a layer of longitudinal muscle cells. The relatively rigid cuticle works with the muscles to create a hydroskeleton as nematodes lack circumferential muscles. Projections run from the inner surface of muscle cells towards the nerve cords; this is a unique arrangement in the animal kingdom, in which nerve cells normally extend fibres into the muscles rather than *vice versa*.

Digestive System

The oral cavity is lined with cuticle, which is often strengthened with ridges or other structures, and, especially in carnivorous species, may bear a number of teeth. The mouth often includes a sharp stylet, which the animal can thrust into its prey. In some species, the stylet is hollow, and can be used to suck liquids from plants or animals.

The oral cavity opens into a muscular, sucking pharynx, also lined with cuticle. Digestive glands are found in this region of the gut, producing enzymes that start to break down the food. In stylet-bearing species, these may even be injected into the prey.

There is no stomach, with the pharynx connecting directly to a muscleless intestine that forms the main length of the gut. This produces further enzymes, and also absorbs nutrients through its single cell thick lining. The last portion of the intestine is lined by cuticle, forming a rectum, which expels waste through the anus just below and in front of the tip of the tail. Movement of food through the digestive system is the result of body movements of the worm. The intestine has valves or sphincters at either end to help control the movement of food through the body.

Excretory System

Nitrogenous waste is excreted in the form of ammonia through the body wall, and is not associated with any specific organs. However, the structures for excreting salt to maintain osmoregulation are typically more complex.

In many marine nematodes, one or two unicellular 'renette glands' excrete salt through a pore on the underside of the animal, close to the pharynx. In most other nematodes, these specialised cells have been replaced by an organ consisting of two parallel ducts connected by a single transverse duct. This transverse duct opens into a common canal that runs to the excretory pore.

Nervous System

Four peripheral nerves run the length of the body on the dorsal, ventral, and lateral surfaces. Each nerve lies within a cord of connective tissue lying beneath the cuticle and between the muscle cells. The ventral nerve is the largest, and has a double structure forward of the excretory pore. The dorsal nerve is responsible for motor control, while the lateral nerves are sensory, and the ventral combines both functions.

At the anterior end of the animal, the nerves branch from a dense, circular nerve ring surrounding the pharynx, and serving as the brain. Smaller nerves run forward from the ring to supply the sensory organs of the head.

The bodies of nematodes are covered in numerous sensory bristles and papillae that together provide a sense of touch. Behind the sensory bristles on the head lie two small pits, or 'amphids'. These are well supplied with nerve cells, and are probably chemoreception organs. A few aquatic nematodes possess what appear to be pigmented eye-spots, but is unclear whether or not these are actually sensory in nature.

Reproduction

Most nematode species are dioecious, with separate male and female individuals. Both sexes possess one or two tubular gonads. In males, the sperm are produced at the end of the gonad, and migrate along its length as they mature. The testes each open into a relatively wide sperm duct and then into a glandular and muscular ejaculatory duct associated with the cloaca. In females, the ovaries each open into an oviduct and then a glandular uterus. The uteri both open into a common vagina, usually located in the middle of the ventral surface.

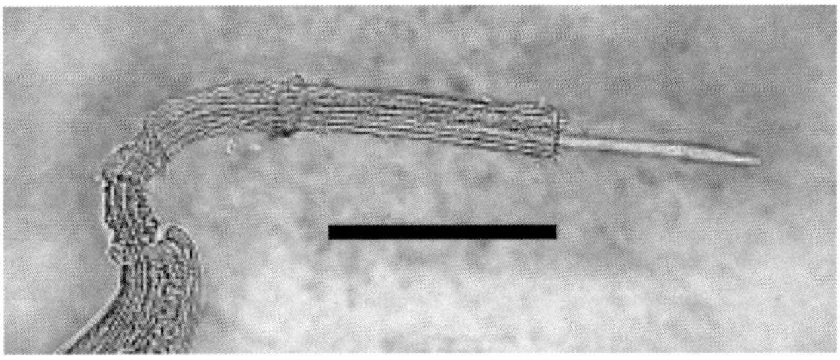

Figure: *Extremity of a male nematode showing the spicule, used for copulation. Bar = 100 μm*

Reproduction is usually sexual. Males are usually smaller than females (often much smaller) and often have a characteristically bent tail for holding the female. During copulation, one or more chitinized spicules move out of the cloaca and are inserted into the genital pore of the female. Amoeboid sperm crawl along the spicule into the female worm. Nematode sperm is thought to be the only eukaryotic cell without the globular protein G-actin.

Eggs may be embryonated or unembryonated when passed by the female, meaning their fertilized eggs may not yet be developed. A few species are known to be ovoviviparous. The eggs are protected by an outer shell, secreted by the uterus. In free-living roundworms, the eggs hatch into larvae, which appear essentially identical to the adults, except for an underdeveloped reproductive system; in parasitic roundworms, the life cycle is often much more complicated.

Nematodes as a whole possess a wide range of modes of reproduction. Some nematodes, such as *Heterorhabditis* spp., undergo a process called *endotokia matricida*: intrauterine birth causing maternal death. Some nematodes are hermaphroditic, and keep their self-fertilized eggs inside the uterus until they hatch. The juvenile nematodes will then ingest the parent nematode. This process is significantly promoted in environments with a low food supply.

The nematode model species *Caenorhabditis elegans* and *C. briggsae* exhibit androdioecy, which is very rare among animals. The single genus *Meloidogyne* (root-knot nematodes) exhibit a range of reproductive modes, including sexual reproduction, facultative sexuality (in which most, but not all, generations reproduce asexually), and both meiotic and mitotic parthenogenesis.

The genus *Mesorhabditis* exhibits an unusual form of parthenogenesis, in which sperm-producing males copulate with females, but the sperm do not fuse with the ovum. Contact with the sperm is essential for the ovum to begin dividing, but because there is no fusion of the cells, the male contributes no genetic material to the offspring, which are essentially clones of the female.

Free-Living Species

In free-living species, development usually consists of four molts of the cuticle during growth. Different species feed on materials as varied as algae, fungi, small animals, fecal matter, dead organisms and living tissues. Free-living marine nematodes are important and abundant members of the meiobenthos. They play an important role in the decomposition process, aid in recycling of nutrients in marine environments, and are sensitive to changes in the environment caused by pollution. One roundworm of note, Caenorhabditis elegans, lives in the soil and has found much use as a model organism. C. elegans has had its entire genome sequenced, as well as the developmental fate of every cell determined, and every neuron mapped.

Parasitic Species

*Nematodes commonly parasitic on humans include ascarids (*Ascaris*), filarias, hookworms, pinworms (*Enterobius*) and whipworms (*Trichuris trichiura*). The species* Trichinella spiralis, *commonly known as the 'trichina worm', occurs in rats, pigs, and humans, and is responsible for the disease trichinosis.* Baylisascaris *usually infests wild animals, but can be deadly to humans, as well.* Dirofilaria immitis *are known for causing heartworm disease by inhabiting the hearts, arteries, and lungs of dogs and some cats.* Haemonchus contortus *is one of the most abundant infectious agents in sheep around the world, causing great economic damage to sheep. In contrast, entomopathogenic nematodes parasitize insects and are considered by humans to be beneficial.*

One form of nematode is entirely dependent upon fig wasps, which are the sole source of fig fertilization. They prey upon the wasps, riding them from the ripe fig of the wasp's birth to the fig flower of its death, where they kill the wasp, and their offspring await the birth of the next generation of wasps as the fig ripens.

A newly discovered parasitic tetradonematid nematode, Myrmeconema neotropicum, *apparently induces fruit mimicry in the tropical ant* Cephalotes atratus. *Infected ants develop bright red gasters (abdomens), tend to be more sluggish, and walk with their gasters in a conspicuous elevated position. These changes likely cause frugivorous birds to confuse the infected ants for berries, and eat them. Parasite eggs passed in the bird's feces are subsequently collected by foraging* Cephalotes atratus *and are fed to their larvae, thus completing the life cycle of* M. neotropicum.

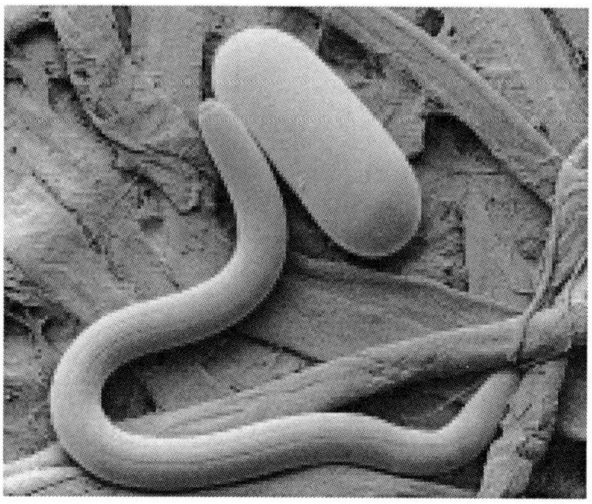

Figure: *Colourized electron micrograph of soybean cyst nematode (*Heterodera *sp.) and egg*

Plant-parasitic nematodes include several groups causing severe crop losses. The most common genera are *Aphelenchoides* (foliar nematodes), *Ditylenchus*, *Globodera* (potato cyst nematodes), *Heterodera* (soybean cyst nematodes), *Longidorus*, *Meloidogyne* (root-knot nematodes), *Nacobbus*, *Pratylenchus* (lesion nematodes), *Trichodorus* and *Xiphinema* (dagger nematodes). Several phytoparasitic nematode species cause histological damages to roots, including the formation of visible galls (e.g. by root-knot nematodes), which are useful characters for their diagnostic in the field. Some nematode species transmit plant viruses through their feeding activity on roots. One of them is *Xiphinema index*, vector of grapevine fanleaf virus, an important disease of grapes, another one is *Xiphinema diversicaudatum*, vector of arabis mosaic virus.

Other nematodes attack bark and forest trees. The most important representative of this group is *Bursaphelenchus xylophilus*, the pine wood nematode, present in Asia and America and recently discovered in Europe.

Agriculture and Horticulture

Depending on the species, a nematode may be beneficial or detrimental to plant health. From agricultural and horticulture perspectives, the two categories of nematodes are the predatory ones, which will kill garden pests like cutworms and corn earworm moths, and the pest nematodes, like the root-knot nematode, which attack plants, and those that act as vectors spreading plant viruses between crop plants. Predatory nematodes can be bred by soaking a specific recipe of leaves and other detritus in water, in a dark, cool place, and can even be purchased as an organic form of pest control.

Rotations of plants with nematode-resistant species or varieties is one means of managing parasitic nematode infestations. For example, marigolds, grown over one or more seasons (the effect is cumulative), can be used to control nematodes. Another is treatment with natural antagonists such as the fungus *Gliocladium roseum*. Chitosan, a natural biocontrol, elicits plant defence responses to destroy parasitic cyst nematodes on roots of soybean, corn, sugar beet, potato and tomato crops without harming beneficial nematodes in the soil. Furthermore, soil steaming is an efficient method to kill nematodes before planting the crop, but indiscriminately eliminates both harmful and beneficial ones.

The Golden Nematode is a particularly harmful variety of nematode pest that has resulted in quaratines and crop failures worldwide.

CSIRO has found a 13- to 14-fold reduction of nematode population densities in plots having Indian mustard (*Brassica juncea*) green manure or seed meal in the soil.

Hundreds of *Caenorhabditis elegans* were featured in a research project on NASA's STS-107 space mission, and were the only known living organisms to have survived the Space Shuttle Columbia disaster.

Globodera Rostochiensis

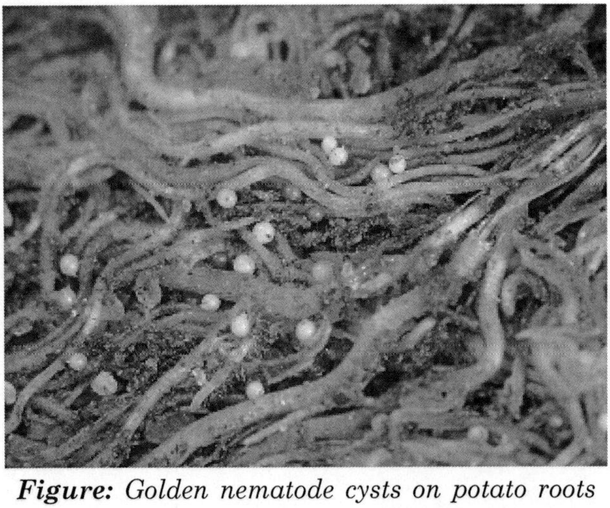

Figure: *Golden nematode cysts on potato roots*

Globodera rostochiensis, commonly known as the golden nematode, golden eelworm or yellow potato cyst nematode, is a plant pathogenic nematode. It is a pest of plants in the family Solanaceae, primarily infesting potatoes and tomatoes, as well as a variety of other root crops.

Pathology

The golden nematode negatively affects plants of the family Solanaceae by forming cysts on the roots of susceptible species. The cysts, which are composed of dead nematodes, are formed to protect the female's eggs and are typically yellow-brown in colour. The first symptoms of infestation are typically poor plant growth, chlorosis, and wilting. Heavy infestations can lead to reduced root systems, water stress, and nutrient deficiencies, while indirect effects of an infestation include premature senescence and increased susceptibility to fungal infections. Symptoms of golden nematode infestation are not unique, and thus identification of the pest is usually performed through testing of soil samples. An infestation may take several years to develop, and can often go unnoticed for between five and seven years. After detection, however, it may take up to thirty years for the pest to be effectively eradicated.

Distribution

The golden nematode, along with the pale cyst nematode, originated in the Andes Mountains of South America. It was first discovered in Germany in 1913, although it is thought to have arrived in Europe with imported potatoes sometime during 19th century. It was first discovered in the United States in 1941, in Canada during the 1960s, and in Mexico during the 1970s. It has also been found in various locations throughout Asia, Africa, and Australia.

The Golden Nematode Blight on the Saanich Peninsula

In the 1960s, the Golden Nematode was discovered in fields in Saanich, BC, Canada. The response of local and national officials provides an interesting insight into the impact a nematode infestation can have on a community. The golden nematode survives best in climates with relatively moist summers and mild winters. Accordingly, while much of Canada's climate remains too harsh for the pest to flourish, the mild climate of southern Vancouver Island is particularly well suited to a golden nematode infestation. In 1965, Central Saanich became the second area in Canada to be quarantined after the discovery of the golden nematode on a local farm.

The presence of the golden nematode resulted in what biologist John Webster deemed a "bombshell." Almost immediately, the seed potato industry and much of the rest of the agricultural export industry throughout British Columbia was shut down. Specialists from Agriculture Canada surveyed all fields in BC in which potatoes were grown in an attempt to detect and isolate the problem. By 1966, it had been determined that the golden nematode was confined to an approximately 150 acre area of land on the Peninsula. It has been suggested that the pest was transmitted to this relatively sheltered area via plant material imported from Europe, possibly in the 1950s. Agriculture Canada quickly took steps to contain and eradicate the nematode. First, a quarantine zone of approximately 450 acres was established. Farmers were provided with a small financial compensation to help them manage during the proposed eradication process. Following the quarantine, infested lands were fumigated with a pesticide called dichloropropene dichloropropane at least twice.

By 1980, however, it was clear that attempts to eradicate the nematode had been unsuccessful. As a result, in 1982, the Department of Agriculture enacted the Golden Nematode Order which banned the growing of all nursery stock, root crops, and other plants which might spread the nematode and restricted the movement of any soil and material from the quarantine area.

Controversy

Financial compensation was arranged for affected farmers. Nevertheless, for many Saanich farmers the quarantine caused considerable and long-lasting financial damage, particularly for those who had invested heavily in formerly lucrative potato industry. In 1991, farmers announced that alternative crops had not been viable replacements for their former crops. Exacerbated by a poor growing season in which some Peninsula farmers lost approximately 40% of their crops, several farms were going into debt while selling their new vegetable crops at a loss. Several farmers called for a revision or repeal of the quarantine, suggesting that the nematode had been effectively eradicated. Lobby groups throughout BC, however, appealed to the provincial and federal governments to retain the ban, arguing that even the remote possibility of spreading the Golden Nematode by opening up the Saanich Peninsula to international trade would negatively impact the BC vegetable industry A decision was made to retain the ban, and appeals to the government by Saanich farmers for redress were refused on the grounds of previous financial restitution.

Impact

The Golden Nematode infestation and the resultant ban on root crop growth, especially potatoes, had both direct and indirect consequences for the Saanich Peninsula community. Foremost, several large-scale farms were forced to sell portions of their land, encouraging a trend away from industrial farming and toward smaller-scale farming on the Peninsula. Additionally, the Peninsula farmers were forced to look to alternate crops and agri-tourism solutions to remain afloat; the Golden Nematode Order banned the growth of potatoes, tomatoes, and peppers, which together account for approximately 50% of all fresh vegetables consumed by Canadians. Without access to this substantial portion of the market, Peninsula farmers sought alternate crops, which included kiwis, berries, and grapes. Thus, combined with outside competition and new industries taking priority, the farming industry has substantially diminished on the Saanich Peninsula. It has been suggested that while fifty years ago, over 90% of food on the Peninsula was grown locally, by 2004 this percentage had dropped below 10%.

Vegetable Plants Nematodes

Nematodes are not insects but microscopic, long, thin worms, which is why a common name for them is eelworms. This soil-borne pest causes stunted, unproductive plants. A common way to identify the problem is infected plants will wilt rapidly in hot weather. When nematodes burrow into the roots they stimulate the development of galls on the roots which become swollen, disfigured and knotty. Root knot nematodes infest a wide range of plants, including roses, potatoes, carrots, tomatoes, cucumbers, lettuces and zucchini.

This pest can be a serious problem for most vegetables, causing infected plants to appear stunted and pale, drop flowers and fruits, wilt often, and decline even when plants are generously watered and fertilized. Gardeners most often realise they have root knot nematode at the end of the season, when they are pulling up spent crops and notice multiple bumpy, knot-like swellings on the roots of vegetable plants. There is nothing available that will kill root knot nematodes, but they can be managed to keep levels low enough to successfully grow most vegetables.

One of the easiest ways to reduce nematode levels is to grow crops that are not susceptible to attack. These include sweet corn, asparagus, and cool season crops in the cabbage family, such as broccoli, kale, collards, and mustard. For some crops that are susceptible to nematode attack, resistant varieties are available.

For many other crops resistant varieties are not available. To grow these crops in nematode infested soils gardeners have to rely on other practices to manage nematode levels. A practice gaining in popularity is the use of certain cover crops to reduce nematode populations. One of the most promising is rapeseed, a relative of mustard and canola. When tilled into the soil, decaying leaves from this crop suppress root knot nematode levels.

Rapeseed is seeded in the fall, from late September through late October. Seed should be broadcast across the garden. Plants are left to grow through the winter and tilled into the soil in March. Rapeseed crops have a high sulfur requirement, a nutrient that is commonly deficient in sandy soils. When growing rapeseed for nematode control, be sure to take a sample of your soil to your local Cooperative Extension office for testing to find out if you need to add additional sulfur. If sulfur levels are too low, the rapeseed crop will not be able to generate the organic compounds that suppress nematode populations.

Other practices that reduce root knot nematode levels in vegetable gardens include frequently tilling the soil in the spring and summer to expose nematodes to sun and air, adding compost to the soil, and soil solarization. Gardeners can solarize their soil by tilling and watering the garden then covering it with clear plastic for several weeks in summer. To keep nematode levels down, gardeners will need to employ as many of these practices as possible every season.

Other plants root knot nematodes will also attack include figs, peaches, gardenia, aucuba, Japanese holly, Japanese boxwood, roses, and dogwoods. It is much more difficult to manage nematodes around permanent plantings and often the best option for landscape beds that have root knot nematodes is to remove infected plants and replace them with species that are resistant.

Suggested Organic Strategies

- A healthy soil will have a range of microorganisms that are predators or parasites of nematodes. All the usual ways of building a healthy soil such as adding compost, mulching and green manuring will improve nematode control.
- A green manure can be grown specifically for nematode control as there are plants that when chopped through the soil will decompose and release a nematode killing gas; this process is known as bio-fumigation. Plants that contain high levels of bio-fumigant compounds include: rapeseed (canola) *Brassica napus*, BQ Mulch, marigolds and Indian mustard. BQ Mulch, canola and mustard are cool season crops. Marigold is a warm season crop that when mixed with cowpea makes an effective, warm season, nematode-controlling green manure.
- Practising crop rotation helps as when a non-host crop is grown for a season it can starve the existing nematodes. Non-host plants include cowpea, soybean, oats, wheat and woolly pod vetch. A rotation of 2 years or more between susceptible crops

is needed to control a serious outbreak. Vegetable crops resistant to nematodes include broccoli, corn, brussels sprouts, chives and leeks.

- Good hygiene will help limit the spread of this pest as nematodes cannot move quickly through the garden, instead they are often spread on infected plants, muddy boots and garden tools.
- When harvesting infected plants, remove as much infected root from the soil as possible and dispose of well away from garden areas. The infected roots can be used as mulch under native shrubs or trees but do not place them in a compost heap, as it is unlikely to get hot enough to kill the nematodes.
- Solarisation can be a useful remedy for nematodes; it can also help combat stubborn weeds. To be effective do this in summer and first water the soil well. Then cover the soil with clear 4mm thick plastic. Stretch the plastic over the area, get it as close to the soil as possible. Bury the edges by digging a narrow trench, tucking the plastic in and back-filling. The aim is to raise the temperature to between 45°C and 50°C in the top 10 cm of soil. This is high enough to kill disease pathogens but most beneficial soil organisms will survive. Leave the plastic in place for 4 to 6 weeks and then plant as usual.
- Digging fresh chicken manure into a hot, dry soil, something normally to be avoided, has been shown to reduce nematode numbers.
- Drenching with water and molasses or sugar can also kill nematodes, but will have a negative impact on soil life.

Not all nematodes are a problem, a range of beneficial nematodes known as 'entomopathogenic' are used to control plant pests.

9

Vegetables Pest and Mites

Correctly identifying the insects and other pests that attack vegetables is the first step toward controlling these pests effectively. The pest descriptions in this booklet are designed to help the commercial vegetable grower and home gardener recognise the insects found on vegetables. Information on insect life cycles and habits is included to make pest control recommendations more understandable and usable. More specific descriptions of insects and detailed information on control procedures are available through local county Agricultural Extension Services offices.

Most of the insects considered common vegetable pests undergo a developmental process known as metamorphosis, which simply means that the insect changes form during its life. Metamorphosis may be complete or incomplete. Complete metamorphosis consists of four stages — egg, larva, pupa, and adult.

All insect life cycles include an egg stage. The insect is immobile during this stage. A larva hatches from the egg. The larval stage is the growth stage and frequently the pest stage of insect life. This is especially true of moths, butterflies, and true flies. Larvae of moths and butterflies are commonly known as caterpillars. Maggots are the larvae of true flies. Many beetles exist as grubs during the larval stage. Larvae vary in size from as short as 1/8 inch to as long as 4 inches.

The larva molts, or changes and enters the pupal stage. As during the egg stage, the insect is immobile during this pupal or changeover period. It is during this stage that the insect develops into an adult.

Adults may or may not feed in the same way and on the same plants as larvae. Beetles, for examplc, have chewing mouthparts and, like many grubs (beetle larvae), are capable of causing great damage to vegetables. Moths and butterflies, on the other hand, have mouthparts designed for siphoning and are unable to damage plants.

Incomplete metamophorsis consists of three developmental stages — egg, nymph, and adult. The egg stage is similar whether an insect undergoes complete or incomplete metamorphosis.

Immature insects called nymphs hatch from eggs. Nymphs damage plants and reduce vegetable quality by sucking plant juices. Some insects damage plants not only by extracting juice but also by injecting saliva into the plant. The best known example of this is the injury called hopperburn caused by the potato leafhopper on Irish potatoes.

The pests discussed in this booklet are divided into the following groups: Moths, Butterflies, and Their Young (Caterpillars); Beetles; Sap-Feeding Insects (True Bugs, Aphids, Leafhoppers, and Whiteflies); Flies; Other Insects; and Noninsect Pests.

Moths, Butterflies, and Their Young

Caterpillars, the larval stage of moths and butterflies, damage both the foliage and fruit of a number of vegetables. These insects chew holes in foliage and fruit and leave degrading excrement and silk on plants.

Cabbage Looper

This caterpillar feeds on the underside of leaves of cabbage and related crops and on lettuce, spinach, beets, peas, and tomatoes. It chews irregular holes through leaves. Moths lay greenish-white eggs on the upper surface of leaves. Larvae, or worms, are 1-1/2 inches long when full grown. Larvae are pale green with four thin, white lines along the back and a wide, pale line on each side of the body. These lines can be seen easily, even on small larvae.

The head is narrower than the thorax (body sextion behind the head). The larvae move in a measuring or looping motion, thus the insect's name. When disturbed, loopers rear upwards. Larvae go into a resting or pupal stage in a flimsy, lace-like cocoon on the plant. The adult is a brown moth with silver markings on the wings. It has a wingspread of about 1-1/2 inches. There may be three or more generations per year, the number of worms increasing greatly with each generation. Cabbage loopers do not overwinter in North Carolina. Moths from Georgia and Florida move into the state each year near the middle of May.

Beet Armyworm

Similar to the true armyworm, this insect is a general feeder and attacks foliage, stems, and sometimes roots of a wide variety of vegetables. Among plants attacked by this pest are beets, asparagus, corn, lettuce, cabbage, tomatoes, peppers, potatoes, onions, and peas. This pest is capable of destroying young plants. The beet armyworm overwinters as a pupa, emerging as an adult in the spring. The adult is a moth with a 1-1/4 inch wingspread. The moth's grayish brown front wings are marked with lighter lines and a spot near the centre of the wing. Hind wings are white with brown edges and veins. Females lay between 500 and 600 eggs over a 4- to 10-day period. Irregular egg masses are covered with scales from the moth. Mature larvae are green with a prominent dark stripe along each side. Larvae have a dark spot on each side of the thorax, are 1-1/4 inches long, and pupate in the soil. In North Carolina, there are several generations a year.

Corn Earworm

This caterpillar feeds on a wide range of vegetables and, as a result, is known by several names. It may attack almost any vegetable but does the most serious damage to corn and tomatoes. Other vegetables attacked include peppers, eggplant, beans, okra, sweet potatoes, lettuce, and cabbage.

Fresh corn silk is preferred by this pest as a location to deposit eggs. Eggs may hatch as soon as two or three days from the time they were deposited. Newly hatched larvae feed on buds and leaves or on corn silks. When they are four or five days old, larvae travel down the silks and feed on kernels at the end of the corn ear. If fresh corn silk is unavailable, moths prefer laying their eggs on tomato foliage, especially when plants are flowering. Larvae attack tomatoes from the time the fruit forms until it ripens, boring into the sides of the fruit and near the stem. Sometimes a small, black entrance hole is the only visible sign that a large worm is inside a tomato. Injury to tomatoes is most severe in the late summer and fall.

When full-grown, larvae are about 1-1/2 inches long. They may be brown, green, or pink, with lighter coloured longitudinal stripes. The head is yellow, and the underside of the body is always lighter than the back. Larvae enter the soil to pupate.

Moths are cream to grayish-brown, have green eyes, and are marked with a dark spot near each wing tip. They have a wingspread of about 1-1/2 inches. Hind wings are light coloured with darker areas near the edges. Moths deposit yellow eggs one at a time on plants. One female can lay as many as 3,000 eggs. There are usually three and sometimes a partial fourth generation of earworms each year in North Carolina.

Cutworm

Many kinds of cutworms attack vegetables. Asparagus, cabbage, squash, and tomatoes are particularly vulnerable. Most cutworms cut off stems of plants at or near the soil. Cutworms feed at night when temperatures are 65 degrees F or higher. During the day these insects hide just beneath the soil close to the site of the previous night's damage. This pest is capable of damage sufficient to necessitate the replanting of corn. Cutworms usually do not eat plants beyond cutting through them, although some species climb stalks and feed on the upper parts of plants, which causes less damage.

When full grown, cutworms are 1-1/2 inches long. They curl up into a tight C shape when disturbed. Most cutworms are a dull colour and have practically no hair. They pupate underground. Cutworms generally overwinter in the soil in the larval or pupal stage. Adults are dull coloured moths with wingspreads of 1 to 1-3/4 inches. Hind wings are usually light in colour. Moths lay eggs on the stems of grasses and weeds or on bare ground. There may be more than one generation of some species each year.

Diamondback Moth

Larvae of this moth feed during the cool parts of the growing season on the underside of leaves of cabbage and related crops. Larvae feed for a time after hatching as leafminers. As larvae grow, they begin feeding on the external portion of leaves and buds. Feeding holes do not penetrate the entire leaf. Larvae are greenish-yellow with black hair and are about 1/3 inch long when full grown. They wriggle actively when disturbed and drop from the plant on silken threads. Larvae pupate inside a lacy cocoon attached to the leaf of the host plant. Moths overwinter in debris of collards, cabbage, cauliflower, and related crops. They have a wingspread of about 1/3 inch. Moths are gray to brown, marked with white. When at rest the moth has white, diamond-shaped markings along the back. The hind wings are paler and have a fringe of hair on the rear margin. Moths lay small, white eggs either singly or in groups of three or four on leaves. There may be five or six generations per year in North Carolina.

European Corn Borer

This borer attacks many species of plants, but corn, peppers, and Irish potatoes are the vegetables most severely damaged. European corn borers feed on leaves and in protected areas until about half-grown, when they bore into a plant's stem or fruit. Nearly an inch long when full grown, the caterpillar is cream coloured, has a brown head, and is marked with small, round, brown spots. Caterpillars molt into pupae inside the host plant. This insect overwinters as a nearly full-grown larvae in cornstalks at ground level. Adults are pale yellow moths with irregular dark bands running across the wings. The moth's wingspread is about 1 inch. Males are smaller and somewhat darker than females. Eggs are laid in groups of up to 50 on the underside of leaves and overlap to resemble small fish scales. Most of North Carolina sees at least two generations of this insect each year, while in the eastern part of the state, there are three generations most years.

Fall Armyworm

Like other armyworms, this insect may "march" from field to field. It often is called the shatterworm because of the destruction it causes to corn foliage. Larvae also damage corn ears as well as beans, potatoes, peppers, tomatoes, spinach, cabbage, and cucumbers. Swarms of moths move northward each spring from southern Florida and areas along the Gulf Coast. Females lay about 1,000 eggs each in masses of about 150. Eggs are usually laid on grass. Larvae are very small when they hatch but reach a length of 1-1/4 inches when full grown. Larvae vary in colour from light tan to green or nearly black, are decorated with thin, yellow stripes, and have a prominent inverted white Y on the front of the head. Larvae go into the soil to pupate. Adult moths have dark gray front wings that are mottled with light and dark splotches. The wingspread of the adult moth is about 1-1/4 inches. Hind wings are grayish-white. Males have a white spot near each wing tip. It appears that this insect cannot overwinter in North Carolina. Most years there are many generations of the pest.

Imported Cabbageworm

This caterpillar attacks cabbage, cauliflower, kale, collards, mustard, radish, and turnips. Larvae riddle outer leaves with irregularly shaped, often large holes. Caterpillars are capable of doing enough damage to severely retard plant growth. Females lay one egg per leaf on the underside of leaves. Eggs hatch into caterpillars that are 1-1/4 inches long when full grown. Caterpillars are velvety-green and are decorated with three indistinct gold stripes, one along the back and one on each side. The pupal or chrysalid stage of the insect is brownish-grey and has many projections or spines. This insect is commonly found attached to a leaf during the pupal stage. Glue at the pupa's tail and a silken loop that encircles the middle of the pupa's body hold it in place. The adult is a white butterfly with one or two black spots on each wing.

This pest overwinters as a pupa either on old plant refuse or on nearby posts or buildings. There are three or four generations per year.

Pickleworm

Pickleworm larvae bore into many cucurbit fruits. Cantaloupes, cucumbers, and squash are most seriously damaged. Caterpillars feed first on foliage. As the insect grows, it begins feeding on fruit. Pickleworms do the greatest damage to growing vine tips and blossoms early in the season. Larvae are initially white with black spots and have dark heads. There is a dark indention on each side on the back of the head. Larger larvae are up to 3/4 inch long and are green. Sawdust-like excrement produced by larvae can easily be seen extruding from damaged fruit. Winter is spent in the pupal stage in a silken cocoon on crop refuse in subtropical regions of the United States. Pickleworms usually appear in North Carolina in early July. The adult is a moth with a wingspread of about 1 inch. Wings appear fragile and are marked with pale yellow centres and yellowish-brown margins.

Potato Tuberworm

This caterpillar does significant damage feeding on exposed potatoes in the field, riddling them with narrow, silk-lined burrows. Adults are small, narrow-winged, grayish-brown moths with a wingspread of 1/2 inch. Moths tend to fly in late afternoon and evening. Females deposit single eggs on foliage of tomatoes, potatoes and related plants, or on exposed tubers of Irish potatoes. Eggs hatch into pinkish-white or green caterpillars with brown heads. When full grown, caterpillars are about 1/2 inch long. Larvae mature in about three weeks and pupate in gray, silken dirt-covered cocoons on the outside of the tuber. A brush of long dark scales protrudes at the end of the body. Moths deposit clusters of two to seven eggs on fruit, vines, buds, and leaves. There are at least four generations per year in North Carolina. Cultural control is possible by planting for early maturity and destroying plants immediately after harvest, thus preventing movement of the pest to later crops.

To prevent damage, potatoes should be kept covered with soil before harvest, and harvested potatoes should be stored in insect-proof areas. Also, vines should be cut, raked, and burned two or three days before harvesting. There may be several generations of this insect per year.

Squash Vine Borer

This borer can be very destructive to squash and pumpkin plants. By boring into a vine near the base, this thick, white, wrinkled brown-headed caterpillar causes entire plants to wilt. The larva overwinters

1 or 2 inches in the soil within a dark, dirt-encased cocoon and molts into a pupa in the spring. Within two or three weeks after molting, the 3/4-inch-long pupa has worked its way to the soil surface and emerged as an adult. The adult squash vine borer is a moth with a wingspread of about 1-1/2 inches. The front wings are covered with metallic, greenish black scales.

Hind wings are transparent. The abdomen is ringed with red and black. Squash vine borer moths look and act like paper wasps. Females lay red eggs near the base of plant stems. Eggs produce smooth, white caterpillars that bore into the centre of the vine a few days after hatching. Small piles of green excrement and wilting of whole vines are signs a plant is infested with squash vine borers. There are two generations of this pest each year. Damage is usually noticed in early June and again in early August.

Tomato Hornworm and Tobacco Hornworm

Tomato and tobacco hornworms feed on tomatoes, peppers, potatoes, and related plants. These green worms are 3 to 4 inches long and can defoliate young plants. Mature tomato hornworms can be distinguished by eight L-shaped, white markings along each side of the body and a bluish-black horn near the rear end of the insect. Tobacco hornworms are decorated with diagonal white bars and have a red horn. The white cocoons of a parasite are usually attached to the worm. Pupae of both species are mahogany brown and are found in the soil. In both species, adults are large, gray moths with mottled front wings. Hind wings are lighter. Two thin, zigzag lines run across the midsection of each hind wing of tomato hornworm moths. There is a dark area between these somewhat obscure lines on the wings of tobacco hornworm moths. On both species, five or six yellowish-orange spots can be found along each side of the body. Moths of both species have a wingspread of 4 to 5 inches and fly mainly around dusk. Females lay eggs individually on plants. Eggs are spherical and greenish-white. This insect overwinters as a pupa. In North Carolina, there are two generations a year.

True Armyworm

This caterpillar is sometimes present in very high numbers. Armyworm epidemics can result in the destruction of large areas of vegetation. This pest attacks grasses, corn, and some legumes. Damage usually starts at the edge of a field. Larvae feed at night and may move in hordes. Armyworms are capable of consuming corn plants up to 8 inches tall. This insect stops feeding and overwinters as a partially grown larva. It resumes feeding in the spring. Armyworms are dark greenish-brown worms up to 2 inches long with white stripes down the sides and back. The insect pupates just below the surface of the soil.

Yellowstriped Armyworm

This caterpillar is a daytime feeder. It is capable of doing significant damage to young tomato plantings, pepper fruit, and, occasionally in August, to sweet potatoes. Its presence is sporadic, however. Overwintering as pupae in the soil, yellowstriped armyworms emerge as moths in the spring. Adults are brown, black, and white moths with a 1-1/2 inch wingspread. The moth's forewings are marked with a diagonal band that runs from the front to the rear of the wing and a smoky white stripe along the outside edge of the wing. Hind wings are white with a dark edge. Eggs, covered with scales from the female's body, are laid in clusters on foliage, trees, and buildings. Larvae are dark green with a yellowish-orange stripe along each side and black spots on their abdominal segments. In North Carolina there are several generations a year.

Beetles

Adult beetles are usually hard-bodied insects with thick forewings. The young are grubs, borers, or wireworms. Often adults feed on different host plants than do larvae, although both stages may be destructive to vegetables.

Asparagus Beetle

There are two beetle species that feed on asparagus — the asparagus beetle and the spotted asparagus beetle. The spotted asparagus beetle, which is reddish-orange and black, is more destructive in northern asparagus production areas. Asparagus beetles are more troublesome in North Carolina. The asparagus beetle is metallic blue with six yellow areas on the wing covers. Beetles are about 1/4 inch long and have a narrow, red head and thorax. This pest overwinters in protected areas and first appears each year in early April along the edges of fields. Adults gnaw tender asparagus buds and lay eggs on developing spears. Eggs are very dark, bulletlike, and laid in rows.

Processing plants sometimes reject otherwise edible spears because eggs have been laid on them. Eggs hatch within a week of the time they were laid. Sluggish, humpbacked, slate-coloured larvae with black heads feed and grow on asparagus ferns for about 10 days. The pupal stage is passed on the soil. A week or so after pupation, adults emerge and the cycle is repeated. There may be as many as five generations in a year. Asparagus is the only vegetable this insect attacks.

Bean Leaf Beetle

Adult bean leaf beetles range in colour from red to yellow. The insect's markings may also vary, but bean leaf beetles generally have a dark triangular mark on the front of the wing covers, six black spots in the centre of the back and a black edge around the outside of the wing covers. This insect is about 1/4 inch long and is long-lived. Adults will drop from foliage when disturbed. Bean leaf beetles feed on foliage, making round, regular holes. This pest also attacks the stems of young plants at or slightly below ground level. Females lay eggs in the soil near the base of plants. From the eggs hatch slender, white larvae that feed on the roots of bean plants. Bean leaf beetles spend the winter as adults in trash near bean fields. There is one generation, possibly two, per year. Beans, peas, cowpeas, and soybeans are common host plants; however, this pest will also feed on corn, some clovers, and many weeds.

Blister Beetle

Several species of blister beetle feed on vegetables. The black, striped, and margined blister beetles are most commonly seen in North Carolina. Blister beetles are slender, rather soft insects with a narrow thorax. The tip of the abdomen is exposed beyond the wing covers. Adults are black or gray and sometimes have narrow, yellow or gray stripes or edges on the wings. Adult beetles are ravenous feeders and are capable of destroying plants. The winter is passed as larvae in the soil. Larvae are about 1/3 inch long and yellow. Blister beetle larvae are beneficial in that they feed on grasshopper eggs. The pupal stage also is spent in the soil. Adults emerge from the soil in midsummer and feed in groups for about five weeks. The beetle's body contains an oil called cantharidin that causes blisters if the insect is crushed against tender skin. Yellow eggs are laid in holes in the soil in clusters of 100 or more. It appears there is one generation per year. Blister beetles will attack most types of vegetables but seem to prefer potatoes and tomatoes.

Colorado Potato Beetle

This insect is perhaps best known as the common yellow- and black-striped potato bug. This pest is, however, a beetle and not a true bug. Both adults and larvae feed on the foliage of potatoes. While potatoes are preferred, this pest also attacks tomatoes, eggplant, peppers, and horse nettle. Adults are about 3/8 inch long and 1/4 inch wide and very convex in shape. Black and yellow stripes run lengthwise along the insect's back. Females lay clusters of 20 or more yellowish-organe eggs on the underside of leaves. A single female beetle can lay over 500 eggs over the course of her four-week life. From the eggs come red, hump-backed, soft-bodied larvae decorated with two rows of black spots along each side of the body. Larvae begin chewing on leaves immediately after hatching. Larvae also feed on the fruit of tomatoes and eggplant. In two or three weeks larvae grow to a length of 1/2 inch, then enter the soil to pupate. Seven days later adult beetles emerge from the soil, feed for a few days, mate, and lay more eggs. This insect overwinters as an adult in the soil. There are at least two generations per year.

Cowpea Curculio

This insect most commonly attacks blackeyed and crowder peas. Adults are black, humpbacked snout beetles about 1/4 inch long. The adult female chews a hole in a pea pod and inserts an egg into the developing seed. This small puncture is difficult to see, and damage may go unnoticed until the crop is harvested. Eggs hatch into small, white, legless grubs with yellowish-brown heads. The grubs destroy the peas in a pod. There is at least one generation per year. Hot, dry weather appears to retard development of this pest. Adults overwinter in refuse around fields. Damage may be avoided by alternating from year to year the site where peas are planted.

Flea Beetle

The many different species of flea beetle attack a variety of vegetables. Species that attack potatoes, tomatoes, and eggplant are the most troublesome, but flea beetles also damage corn and sweet potatoes. Flea beetles are so named because of their enlarged hind legs and jumping ability. Adults are 1/16 to 1/18 inch long and usually black with brown legs and antennae. Eggs are laid in the soil at the base of plants. From the eggs hatch cylindrical, brown-headed, white larvae that are about 1/5 inch long when full grown. Larvae commonly feed on roots, often riddling them with tunnels or eating off rootlets. The pupal stage is spent in the soil. These insects overwinter as adults in trash around field borders and in ditch banks. There are three or more generations per year. Round holes chewed in leaves are the most obvious damage caused by flea beetles. On small plants, this "buckshot" damage can result in plant death. On larger plants these feeding sites provide a good entrance for several blights and wilts.

Japanese Beetle

These beetles are metallic green or greenish-bronze, 1/2 inch long with red wing covers, and are decorated with white spots near the tip of the abdomen. They commonly are found in June and July on fruit, deciduous trees, and vegetables. Japanese beetles also are frequent and persistent pests of sweet corn and beans. The insect spends the winter in the soil as a white grub about 3/4 inch long. Beetles emerge from the soil in midsummer and on warm, sunny days swarm to plants, where they eat the foliage, flowers, and fruit. Females beetles then return to grassy areas to lay their white, spherical eggs in the soil. Grubs feed on decaying vegetation and grass roots. Pupation occurs in the soil. There is only one generation each year.

Mexican Bean Beetle

The brown Mexican bean beetle is decorated with 16 black spots arranged in three rows across its back. Beetles are 1/3 inch long, convex, and resemble lady beetles. Adults overwinter on the ground among leaves and trash near fields where beans were grown. When warm weather arrives in the spring, they leave their overwintering sites and feed on beans for a week or two. Females lay lemon-coloured eggs in groups of 25 or more on the underside of leaves. Eggs hatch about 10 days after being laid and yellow larvae with six rows of black-tipped, branched spines on their backs appear. Larvae, which are 1/3 inch long when full grown, feed for two to five weeks, then enter the pupal stage. The yellow pupa is attached to a leaf. Traces of the last larval skin can usually be found where the pupa is attached to the leaf. There are three or four generations per year. Both larvae and adults feed on leaves, usually on the underside. Damaged leaves have a skeletonized or lace-like appearance. Sometimes stems and pods are damaged. The insect attacks all kinds of garden beans, cowpeas, and soybeans.

Spotted Cucumber Beetles

In its adult form, this insect feeds on over 200 different plants. It may be found on beans, cucurbits, corn, potatoes, asparaugs, cabbage, peas, beets, tomatoes, and turnips. It feeds on foliage and new shoots, leaving small holes in foliage. It also gnaws holes in fruit. This insect overwinters as a 1/4 inch long, yellowish-green beetle decorated with 12 black spots. The beetle's head and antennae are black. Females deposit eggs at the base of grass and corn plants. From the eggs hatch yellowish-white, brown-headed larvae that are up to 3/4 inch long when full grown. These larvae, called corn rootworms, tunnel in the roots of plants. This tunneling action can result in severe stunting of corn plants and sometimes kills plants. There appears to be two generations per year in North Carolina.

Striped Cucumber Beetle

These 1/5-inch long beetles are yellow and are decorated with three wide black stripes. Striped cucumber beetles damage plants by feeding on leaves, stems, and fruit. They also carry the organisms that cause bacterial wilt of cucurbits and cucumber mosaic. Adults lay yellow eggs at the base of plants. Larvae, which are white and 1/3 inch long, feed on the underground parts of plants or on fruit that touches the soil. There are at least two generations per year. In addition to cucurbits, striped cucumber beetles feed on beans, peas, and corn. The most severe damage is caused to small plants.

Vegetable Weevil

Larvae and adults of this insect feed on foliage and exposed roots of potatoes, turnips, turnip greens, kale, and mustard greens. Late fall and early spring crops are most severely damaged. Adults are gray snout beetles 1/3 to 1/2 inch long. Two white markings on the rear half of the wing covers form a V shape. Eggs laid in the crowns of plants procude green, slug-like larvae. These grubs are about 1/2 inch long when full grown. This pest goes into a resting stage from June to September.

Wireworms

These beetles do the most severe damage to root crops such as Irish and sweet potatoes, although they injure other vegetables as well. There are many different species of wireworms. Adults are hard-shelled, dull coloured beetles, ranging from 1/2 to 1-1/2 inches in length. Beetles are tapered toward both ends and appear streamlined. Adults are also known as click beetles because they snap and flip over when turned on their backs. They have a loose, flexible joint just ahead of the wings. Females lay eggs in the soil around the roots of grass plants. The eggs produce yellow to reddish-brown, tough-bodied larvae that are commonly called wireworms. Larvae range in size from 1/2 to 2 inches when full grown, and most are cylindrical and slender. Some wireworms are almost flat. Larvae feed on underground parts of plants. The pupal stage is spent in the soil. It takes over a year to complete the life cycle of a wireworm. Wireworm damage is most severe on land not previously planted in row crops.

White Grubs

There are several species of these pests. Adults are hard-shelled, light brown to almost black beetles from 3/4 to 1 inch long. These insects are sometimes called May beetles and June beetles. They commonly feed at night on the foliage of trees and hide in the grass and lay eggs during the day. The spherical, pearly white eggs produce white C-shaped grubs with brown heads. Grubs feed on grass roots and other crops. The pupal stage is spent in the soil. Among the common white grubs are the larvae of the green June beetle. Larvae feed mainly on organic matter in the soil. Grubs uproot young plants as they move about in and on the soil. Vegetables should not be planted in soil that was in sod the previous year until the soil has been treated to control white grubs as some grubs live two to three years.

Whitefringed Beetle

Among the many vegetables attacked by whitefringed beetles are Irish potatoes, sweet potatoes, beans, and cabbage. Adults, which emerge in June and July, are dark gray beetles about 1/2 inch long, with a white margin or fringe on their sides. White, oval eggs are laid in a gelatinous mass on objects on or near the ground. Grubs are shite, legless, and 1/2 inch long when full grown. Grubs feed on the underground parts of many kinds of plants. The pupal stage is spent in the soil, while the insect overwinters in the grub stage. There is one generation per year.

Sap-Feeding Insects

Some insects damage vegetables by sucking plant juices from leaves, stems, or roots. The true bugs, aphids, whiteflies, and leafhoppers, all have beak-like mouthparts that are used to suck sap from plants. Some of these insects also produce toxic saliva, which they inject into plants. The best known example of plant injury caused by saliva injection is a condition called hopperburn caused by the potato leafhopper on Irish potatoes. Many of these insects also transmit plant diseases. The green peach aphid, for example, is known to transmit potato leafroll virus, while the melon aphid infects squash and watermelon mosaic virus.

Harlequin Bug

These bugs feed on cabbage and related crops and can cause severe damage. Harlequin bugs also feed on beans, asparagus, okra, and tomatoes. The insect damages a plant by extracting juices from it. Adults and nymphs have similar habits. Adults are flat, shield-shaped bugs about 3/8 inch long. They are red or organe with black markings that give the appearance of a mask. Eggs resemble kegs and are laid on end in groups. Each egg has a black band at the top and bottom. Nymphs, which grow through five developmental stages, resemble adults but are smaller and wingless. There are two or more generations of this insect per year. Adults overwinter in any protected area.

Eggplant Lace Bug

Adult eggplant lace bugs are small, flat insects about 1/8 inch long. They are somewhat rectangularly shaped when viewed from above. The wings are net or lace-like. The head is covered with a hood-like structure.

Females lay eggs on the underside of leaves. Nymphs are darker than the adults, wingless, and usually covered with spines. Both nymphs and adults inject saliva and suck juices from eggplant leaves. The insect also excretes a brown, varnish-like material.

Squash Bug

This insect attacks only cucurbits. It prefers squash and pumpkins. Adults and nymphs injure plants by sucking sap from them. Damage usually can be detected by the occasional drooping of a leaf. Adults are about 5/8 inch long, brownish-black or gray, and flat across the back. This insect overwinters as an adult and mates in the spring. Reddish-brown eggs are laid on leaves in rows that come together at an angle. Each cluster contains 15 or more eggs, which produce tiny, light gray, wingless nymphs. There is one generation each year. Destruction of crop residue as soon as the crop is harvested helps to control these insects.

Tarnished Plant Bug

Among the many different vegetables this insect attacks are asparagus, Irish potatoes, chard, celery, and beans. A small spot surrounded by a yellow area is evidence a tarnished plant bug has been feeding on a bean seed. Blossoms may drop from a plant as a result of tarnished plant bugs feeding on them. This insect also feeds on the terminals of potatoes, causing leaves near the top of the plant to wilt. Tarnished plant bugs also transmit disease. Adults are small, flat bugs about 1/4 inch long. These somewhat oval insects are mottled brown with irregular splotches of white, yellow, reddish-brown, and black. Eggs are usually inserted into plant tissues. The nymphs resemble adults but are smaller and wingless. There appear to be four or five generations per year in North Carolina. This insect overwinters as an adult. Two other true bugs — stink bugs and leaffotted bugs — cause similar injury, especially on beans and southern peas. Damage inflicted by these two bugs on southern peas and butter beans occasionally is serious.

Aphids

Aphids, or plant lice, feed on most vegetables. These insects cause the greatest damage when they suck juices from and inject saliva into plants. Many aphid species transmit viruses that cause diseases. Adults are soft-bodied, pear-shaped, and may or may not have wings. Two cornicles, or "exhaust pipes," extend from the hind end of most aphids. Aphids may be green, pink, yellow, or black. Nymphs resembe adults but are smaller and always wingless. These pests usually feed in clusters.

Potato Leafhopper

This insect feeds on many different host plants including beans and Irish potatoes. Saliva injected by this pest into Irish potato plants causes a deadening and upcurling of the tissues known as hopperburn. Sap feeding on beans causes leaves to be stunted, crinkles, and curled downward. Adults are pale green, wedge-shaped, and about 1/8 inch long. Eggs inserted into plant tissues produce green nymphs that resemble adults but are smaller and wingless. Adults and nymphs suck plant juices. Potato leafhoppers are a problem primarily in western North Carolina. At least three generations of this pest occur each year. The potato leafhopper is one of several leafhopper species found on vegetables.

Whitefly

Whiteflies feed on a wide variety of vegetables. This insect can be found in the field and the greenhouse. Host plants include beans, tomatoes, eggplant, and cucumber. Adults are very small sucking insects with two pairs of broadly rounded wings. The wings and body are covered with a snow white, waxy powder. The insects look like tiny moths and may fly up in a cloud when disturbed. Eggs, which are almost microscopic, are laid on the underside of leaves. Young nymphs are small, flat, and hard to see. All stages are found on the underside of leaves. Leaves may become discoloured and fall from a plant on which whiteflies are feeding. Plants on which this insect feeds may also be stunted. Whiteflies secrete honeydew on plants, which allows unsightly black sooty mold to grow.

Flies

True flies are a diverse group. Many species attack humans and animals and can spread diseases. Some flies are beneficial. They feed on decaying matter and prey on or parasitize other insects. Others are plant pests, their maggots or larvae feeding on various vegetables.

Cabbage Maggot

This insect prefers cabbage, cauliflower, broccoli, and related crops. It also attacks beets, radishes, turnips, celery, and other vegetables. The cabbage maggot does most of its damage in the mountain regions of North Carolina. The adult is a gray fly with two wings. It is smaller than a house fly but has proportionately longer legs. Eggs are white and are laid on the soil close to young plants. The eggs hatch into white, wedge-shaped maggots that are about 1/4 inch long when full grown. Maggots feed on roots, either on the surface or by tunneling through them. Fleshy parts of radishes or turnips will show brown streaks where tunneling has occurred. Cabbage maggots overwinter in the pupal stage, which is spent in the soil. There appear to be three generations per year of this pest.

Pepper Maggot

This pest prefers peppers but also attacks eggplant. It is believed that its natural food is horsenettle. Adults are yellow striped flies about 1/4 inch long with dark bars on the wings. Females deposit eggs under the skin of peppers. Maggots are white, legless, and about 1/2 inch long when full grown. Maggots eat the core of the pepper, causing it to decay or drop from the plant. The pupal stage, in which the insect overwinters, is spent buried 2 to 4 inches in the soil. Adult flies appear in late June. There is one generation each year.

Seedcorn Maggot

This insect attacks the sprouting seeds of many different vegetables. Damage may be done to seed pieces of Irish potatoes; seedlings of corn, beans, peas, and cucumbers; or stems of other vegetables. The most serious infestations occur in the spring during cool, wet weather and in soils high in organic matter. Adults are grayish-brown flies about 1/5 inch long. Eggs, from which hatch cream-coloured, wedge-shaped maggots, are deposited in the soil close to plants. Maggots are 1/4 inch long when full grown, legless, and have chewing mouthparts at the narrow end of the body. It is as a maggot that this insect is destructive. This pest overwinters in the soil in the pupal stage. The pupa is enclosed in a brown, capsule-like case. There are three generations of this pest each year. Destruction of crop refuse and shallow planting in warm soils will enhance plant establishment.

Vegetable Leafminer

This insect's name derives from the maggot's habit of tunneling or mining in leaves — feeding between the upper and lower leaf surfaces. Damage appears as winding white trails or broad white spots. Leafminers attack many vegetables but do the most damage to cucumbers, squash, tomatoes, and leafy vegetables such as cabbage, kale, and collards. The leaves of a plant that is attacked by vegetable leafminers are weakened, while the mines serve as sites for disease and decay. Adults are gray to shiny black flies less than 1/8 inch long. Small, white eggs deposited in leaves produce white or yellow maggots that tunnel through the leaves. The tunnel is narrow and winding at first. It becomes larger as the maggot grows, finally forming a blotch. The pupal stage is spent in the soil, and it is in this stage that this pest overwinters. There are several generations each year in North Carolina.

Other Insect Pests

Moths and butterflies, beetles, sap-feeding insects, and flies are not the only insects capable of damaging vegetables. Grasshoppers, mole crickets, and thrips also attack vegetables.

Grasshoppers

Many species of grasshopper damage vegetables. Usually these pests migrate from dried up vegetation to more succulent vegetables. Adults and nymphs cause damage by feeding on foliage. Adults have front wings that are longer than the body and are held roof-like over the insect. Hind wings are softer and are folded fanlike beneath the forewings when the insect is at rest. The hind legs are long and adapted for jumping. Size varies according to species. Nymphs resemble adults in habits and appearance but are smaller and wingless. Most species overwinter as eggs, which are laid in packets in the soil.

Mole Cricket

This insect burrows in the soil, causing major damage by uprooting young plants. Mole crickets are capable of severely damaging the roots and tubers of Irish and sweet potatoes. Damage to foliage may also occur. Adults are velvety brown and about 1-1/2 inches long. This insect has short forewings and long, membranous hind wings. The front legs, rake-like and developed for digging, somewhat resemble the forepaws of a mole. Hind legs are enlarged. Eggs are laid in an underground chamber. Nymphs are small, wingless versions of adults. Nymphs and adults overwinter. There is one generation each year in North Carolina.

Thrips

Several species of thrips damage vegetables. The most severe damage usually occurs in hot, dry years. Damage appears as white flecks or streaks on leaves. Adults are small slender insects about 1/20 inch long. They vary in colour from light to dark brown. Wings are narrow and fringed, giving them a feathery appearance. The males of some species are wingless. Eggs are thrust into plant tissues. Nymphs are active, light-coloured, wingless insects. The mouthparts of adults and nymphs are similar and unusual; they are designed to rasp plant tissues then suck the juices. The last nymphal stage is spent in the soil. Adults and nymphs pass the winter on plant tissue. There may be as many as five generations each year in North Carolina.

Noninsect Vegetable Pests

Spider mites and slugs, although not insects, are capable of inflicting severe damage on vegetables.

Mites, along with ticks, are small arthropods belonging to the subclass Acari (also known as Acarina) and the class Arachnida. The scientific discipline devoted to the study of ticks and mites is called acarology. In soil ecosystems, mites are favoured by high organic matter content and by moist conditions, wherein they actively engage in the fragmentation and mixing of organic matter.

Diversity and Systematics

House hold mites can not be transferred from a house to another house. Mites can only live while inhabiting whatever it is that they are inhabiting. For example, if you have mites in a blanket and you bring it outside they will die due to the sudden change in atmosphere. People with mites either have animals or work in the food industry, or live with someone who is in the food industry. Most mites like to live in the same area with other mites though some might live alone.

Mites are among the most diverse and successful of all the invertebrate groups. They have exploited an incredible array of habitats, and because of their small size (most are microscopic,) go largely unnoticed. Many live freely in the soil or water, but there are also a large number of species that live as parasites on plants, animals, and some that feed on mold. It is estimated that 48,200 species of mites have been described.

Some of the plant pests include the so-called spider mites (family Tetranychidae), thread-footed mites (family Tarsonemidae), and the gall mites (family Eriophyidae). Among the species that attack animals are members of the sarcoptic mange mites (family Sarcoptidae), which burrow under the skin. Demodex mites (family Demodicidae) are parasites that live in or near the hair follicles of mammals, including humans. Acari are mites, except for the three families of ticks. Perhaps the best-known mite, though, is the house dust mite (family Pyroglyphidae).

Insects may also have parasitic mites. Examples are *Varroa destructor*, which attaches to the body of the honeybee, and *Acarapis woodi* (family Tarsonemidae), which lives in the tracheae of honey bees. There are hundreds of species of mites associated with other bee species, and most are poorly described and understood. Some are thought to be parasites, while others beneficial symbionts. Mites also parasitize some ant species, such as *Eciton burchellii*.

The tropical species *Archegozetes longisetosus* is one of the strongest animals in the world, relative to its mass (100 ìg): It lifts up to 1,182 times its own weight, over five times more than would be expected of such a minute animal.

Spider Mites or Red Spiders

Mites are closely related to insects. Many species of this pest attack vegetables. Beans and tomatoes may be severely damaged, while crops such as eggplant, cucumber, and celery are also susceptible. These pests appear to cause more severe damage during hot, dry years. Adults are usually red, yellow, or green, about 1/ 50 inch long, and have four pairs of legs. The body is oval. Adults lay eggs on plants. The eggs hatch into young mites that resemble adults but are smaller. Adults and nymphs suck plant juices. Plants attacked by mites appear sickly and white-speckled. Affected leaves curl, turn brown, and drop from the plant prematurely. When an infestation is severe, fine silk webs may entangle an entire plant, and moving spider mites can easily be seen in the webbing. Many generations may occur each year.

Slugs

Slugs are similar to snails but lack the snail's hard, protective shell. Six species of slugs attack vegetables and ornamental plants in North Carolina. Slugs feed on mold, decaying organic matter, and live plant tissue including foliage, flowers, or fruit. Slugs may chew jagged edges along leaves, shred leaves, or eat completely through leaves or stems. Damage is most severe in damp areas or during extended rainy periods. Slugs vary in size from less than 1/2 inch to more than 4 inches long and range in colour from ash-gray or yellow-gray to black. Some are mottled. Young slugs resemble adults but are smaller and lighter in colour. Eggs, gelatinous and water appearing, are laid in clusters of about 25 on the soil in concealed, moist areas. Slugs are active at night, leaving silvery slime trails that can be seen the next morning.

10

Insecticide and Pesticide

An insecticide is a substance used to kill insects. They include ovicides and larvicides used against insect eggs and larvae, respectively. Insecticides are used in agriculture, medicine, industry and by consumers. Insecticides are claimed to be a major factor behind the increase in agricultural 20th century's productivity. Nearly all insecticides have the potential to significantly alter ecosystems; many are toxic to humans; some concentrate along the food chain.

Insecticides can be classified in different ways:

- *Systemic* insecticides are incorporated by treated plants. Insects ingest the insecticide while feeding on the plants.
- *Contact* insecticides are toxic to insects when brought into direct contact. Efficacy is often related to the quality of pesticide application, with small droplets (such as aerosols) often improving performance.
- *Natural* insecticides, such as nicotine, pyrethrum and neem extracts are made by plants as defences against insects. Nicotine-based insecticides are widely used in the US and Canada, but are barred in the European Union.
- *Plant-incorporated protectants* (PIPs) are systemic insecticides produced by transgenic plants. For instance, a gene that codes for a specific *Baccilus thuringiensis* biocidal protein was introduced into corn and other species. The plant manufactures the protein which kills the insect when consumed.

- *Inorganic* insecticides are contact insecticides that manufactured with metals and include arsenates, copper and fluorine compounds, which are now seldom used, and sulfur, which is commonly used.
- *Organic* insecticides are contact insecticides that comprise the largest numbers of pesticides available for use today.

The *mode of action* describes how the pesticide kills or inactivates a pest. It provides another way of classifying insecticides. Mode of action is important in understanding whether an insecticide will be toxic to unrelated species, such as fish, birds and mammals.

Major Classes

Organochlorides: The best-known organochloride, DDT, was created by Swiss scientist Paul Müller. For this discovery, he was awarded the 1948 Nobel Prize for Physiology or Medicine. DDT was introduced in 1944.

It functions by opening sodium channels in the insect's nerve cells. The contemporaneous rise of the chemical industry facilitated large-scale production of DDT and related chlorinated hydrocarbons.

Organophosphates and Carbamates

Organophosphates are another large class of contact insecticides. These also target the insect's nervous system. Organophosphates interfere with the enzymes acetylcholinesterase and other cholinesterases, disrupting nerve impulses and killing or disabling the insect. Organophosphate insecticides and chemical warfare nerve agents (such as sarin, tabun, soman, and VX) work in the same way. Organophosphates have a cumulative toxic effect to wildlife, so multiple exposures to the chemicals amplifies the toxicity.

In the US, organophosate use declined with the rise of substitutes.

Carbamate insecticides have similar mechanisms to organophosphates, but have a much shorter duration of action and are somewhat less toxic.

Pyrethroids

Pyrethroid pesticides mimic the insecticidal activity of the natural compound pyrethrum. These compounds are nonpersistent sodium channel modulators and are less toxic than organophosphates and carbamates.

Compounds in this group are often applied against household pests.

Neonicotinoids

Neonicotinoids are synthetic analogues of the natural insecticide nicotine (with much lower acute mammalian toxicity and greater field persistence). These chemicals are acetylcholine receptor agonists. They are broad-spectrum systemic insecticides, with rapid action (minutes-hours). They are applied as sprays, drenches, seed and soil treatments. Treated insects exhibit leg tremors, rapid wing motion, stylet withdrawal (aphids), disoriented movement, paralysis and death. Imidacloprid may be the most common. It has recently come under scrutiny for allegedly pernicious effects on honeybees and its potential to increase the susceptibility of rice to planthopper attacks.

Ryanoids

Ryanoids are synthetic analogues with the same mode of action as ryanodine, a naturally occurring insecticide extracted from *Ryania speciosa* (Flacourtiaceae). They bind to calcium channels in cardiac and skeletal muscle, blocking nerve transmission. Only one such insecticide is currently registered, Rynaxypyr, generic name chlorantraniliprole.

Plant-Incorporated Protectants

Transgenic crops that act as insecticides began in 1996 with Bt corn that produces the Cry protein, derived from the bacterium Bacillus thuringiensis, which is toxic to moth larvae such as the European corn borer. The technique has been expanded to include the use of RNA interference RNAi that fatally silences crucial insect genes. RNAi likely evolved as a defence against viruses. Midgut cells in many larvae take up the molecules and help spread the signal. The technology can target only insects that have the silenced sequence, as was demonstrated when a particular RNAi affected only one of four fruit fly species. The technique is expected to replace many other insecticides, which are losing effectiveness due to the spread of pesticide resistance.

Insect Growth Regulators

Insect growth regulator (IGR) is a term coined to include insect hormone mimics and an earlier class of chemicals, the benzoylphenyl ureas, which inhibit chitin(exoskeleton) biosynthesis in insects. Diflubenzuron is a member of the latter class, used primarily to control caterpillars that are pests. The most successful insecticides in this class are the juvenoids (juvenile hormone analogues). Of these, methoprene is most widely used. It has no observable acute toxicity in rats and is approved by World Health Organisation (WHO) for use in drinking water cisterns to combat malaria. Most of its uses are to combat insects where the adult is the pest, including mosquitoes, several fly species, and fleas. Two very similar products, hydroprene and kinoprene, are used for controlling species such as cockroaches and white flies. Methoprene was registered with the EPA in 1975. Virtually no reports of resistance have been filed. A more recent type of IGR is the ecdysone agonist tebufenozide (MIMIC), which is used in forestry and other applications for control of caterpillars, which are far more sensitive to its hormonal effects than other insect orders.

Biological Insecticides

Many plants exude substances to repel insects. Premier examples are substances activated by the enzyme myrosinase. This enzyme converts glucosinolates to various compounds that are toxic to herbivorous insects. One product of this enzyme is allyl isothiocyanate, the pungent ingredient in horseradish sauces.

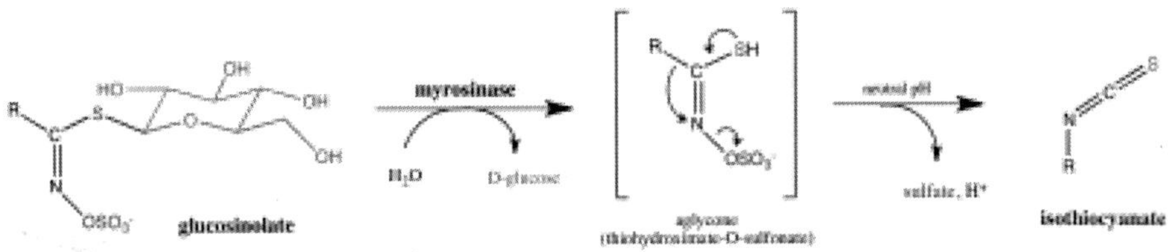

Figure: *Biosynthesis of antifeedants by the action of myrosinase.*

The myrosinase is released only upon crushing the flesh of horseradish. Since allyl isothiocyanate is harmful to the plant as well as the insect, it is stored in the harmless form of the glucosinolate, separate from the myrosinase enzyme.

In general, tree rosin is considered a natural insecticide. To be specific, the production of oleoresin by conifer species is a component of the defence response against insect attack and fungal pathogen infection.

Bacterial Insecticides

Bacillus thuringiensis is a bacterial disease that affects Lepidopterans and some other insects. Toxins produced by strains of this bacterium are used as a larvicide against caterpillars, beetles, and mosquitoes. Toxins from *Saccharopolyspora spinosa* are isolated from fermentations and sold as Spinosad. Because these toxins have little effect on other organisms, they are considered more environmentally friendly than synthetic pesticides. The toxin from *B. thuringiensis* (Bt toxin) has been incorporated directly into plants through the use of genetic engineering. Other biological insecticides include products based on entomopathogenic fungi (e.g., *Beauveria bassiana, Metarhizium anisopliae*), nematodes (e.g., *Steinernema feltiae*) and viruses (e.g., *Cydia pomonella* granulovirus).

Environmental Effects

Effects on Nontarget Species

Some insecticides kill or harm other creatures in addition to those they are intended to kill. For example, birds may be poisoned when they eat food that was recently sprayed with insecticides or when they mistake an insecticide granule on the ground for food and eat it.

Sprayed insecticide may drift from the area to which it is applied and into wildlife areas, especially when it is sprayed aerially.

DDT

The development of DDT was motivated by desire to replace more dangerous or less effective alternatives. DDT was introduced to replace lead and arsenic-based compounds, which were in widespread use in the early 1940s.

DDT was brought to public attention by Rachel Carson's book *Silent Spring*. One side-effect of DDT is to reduce the thickness of shells on the eggs of predatory birds. The shells sometimes become too thin to be viable, reducing bird populations. This occurs with DDT and related compounds due to the process of bioaccumulation, wherein the chemical, due to its stability and fat solubility, accumulates in organisms' fatty tissues. Also, DDT may biomagnify, which causes progressively higher concentrations in the body fat of animals farther up the food chain. The near-worldwide ban on agricultural use of DDT and related chemicals has allowed some of these birds, such as the peregrine falcon, to recover in recent years. A number of organochlorine pesticides have been banned from most uses worldwide. Globally they are controlled via the Stockholm Convention on persistent organic pollutants. These include: aldrin, chlordane, DDT, dieldrin, endrin, heptachlor, mirex and toxaphene.

Pollinator Decline

Insecticides can kill bees and may be a cause of pollinator decline, the loss of bees that pollinate plants, and colony collapse disorder (CCD), in which worker bees from a beehive or Western honey bee colony abruptly disappear. Loss of pollinators means a reduction in crop yields. Sublethal doses of insecticides (i.e. imidacloprid and other neonicotinoids) affect bee foraging behaviour. However, research into the causes of CCD was inconclusive as of June 2007.

Vegetable Garden Insecticides

Even in the best managed vegetable gardens — with soil rich in compost, and a diversity of plant species that encourage natural predators — certain pests will get out of hand.

There is a nugget of truth in the old maxim that insects are most likely to gang up on plants that are already unhealthy because of a soil nutrient imbalance or drought. This is the case, for example, with some aphid outbreaks. Unfortunately, the maxim doesn't come close to explaining all our pest problems. For instance, vegetables are bred largely for yield and flavour, often at the expense of natural resistance to pests. Furthermore, all vegetables are tender and nutritious, and this fact is not lost on a wide array of insects. With cabbageworms, hornworms, bean beetles and Colorado potato beetles, the better you've made the soil, the more they like your vegetables.

If these or similar insects are in your neighbourhood and you are growing their favourite crops, you are almost certain to have a pest outbreak. Given these realities, what should you do? Most likely you will consider using some kind of insecticide.

Here we summarize the latest experience and expert advice about the sprays and dusts used to control pests in vegetable gardens. Also described are two promising new insecticides.

Integrated Pest Management

Vegetable garden pest control begins with basic good gardening common sense, such as choosing varieties that are resistant to pests in your region, preparing the soil well and providing regular irrigation. It helps to have in your garden a diversity of plants and many kinds of habitats. Water, even a very small pond, is very attractive to many insects and other creatures. Likewise, an abundance of flowering, nectar-bearing plants will encourage and sustain parasitic and predatory insects.

The next step in least a toxic-pest control strategy is to employ barriers, such as row covers, to exclude pests altogether. Using a pesticide, any pesticide, is always the measure of last resort. You spray or dust late in the game, when the pest insect is clearly way out of control and an important crop is at risk.

When a pest first arrives, or when prior experience tells you it soon will, the best approach is to develop a strategy of control. Integrated pest management (IPM) is a pest problem-solving process that includes considerations such as pesticide resistance, natural biological controls and pollution in addition to problems caused by the pest. IPM integrates many pest control methods and minimizes insecticide use, particularly the more toxic, broad spectrum kinds.

When a problem does occur, it is essential to correctly identify the cause. The beetle you see near a hole in a leaf may be a predator. But if it is damaging your plants, simply pick it off. Also consider that doing nothing at all — letting nature take its course — is often the best approach. Always use simple, noninvasive remedies first.

Sensible Insecticides Used Responsibly

Sometimes pest problems are not adequately managed by natural, cultural, or mechanical control methods. Insecticides are often the only control option that remains. The prime factors in determining pesticide safety are: how specific it is to particular insects; how toxic it is to humans and nontarget organisms; and how quickly it degrades. Remember, in a vegetable garden it is virtually impossible to spray just the one thing you want to spray. Other crops (perhaps ready to pick) are always nearby, so you want to stay away from insecticides that don't break down quickly.

Choose an insecticide that is as specific to the pest at hand as possible and then use as little as possible. If only one spray will do the job, use only one spray. For the long-term health of your garden, the less spray you use, the better.

Remember, too, that just because an insecticide has a botanical origin or is considered acceptable to organic gardeners, it still contains a toxin and is not automatically safe for humans.

Eleven Vegetable Pest Remedies

Gardeners today have at their disposal a handful of effective and safe pesticides. When you have to spray to save your crop, here are the insecticides to consider using, with their characteristics, positive and negative.

Adios (Sevin Plus Bait).

This environmentally smart new insecticide is aimed at adult cucumber beetles. The active ingredient, Sevin, is inside a microcapsule of bait that cucumber beetles find irresistible. The bait is cucurbitacin, a natural compound in leaves of cucumbers, squash and melons. Once cucumber beetles encounter the Adios pellets on leaves they seek out more, gradually consuming enough Sevin to cause their demise. Very little Sevin is applied, between 2% and 5% the usual amount. Other insects, such as bees and wasps, don't eat the pellets so aren't affected.

And the Sevin does not easily leach out of the pellets into the spray or irrigation water.

BT

The bacterium Bacillus thuringiensis (Bt) was identified in 1911 by the biologist E. Berliner, who found it infecting pupae of the Mediterranean flour moth and other insect larvae living in grain warehouses in the German town of Thuringia. It wasn't until the 1960's, however, that entomologists learned how to make it into a powerful and very pest-specific insecticide.

Advantages of Bt include safety — it is essentially nontoxic to humans, other mammals, and birds. The label specifies no waiting period between application and harvest. It is also highly selective so easily incorporated with existing natural controls. A limitation of Bt is its slow action. After pests consume it, their feeding slows down. But their death won't occur for two to five days. Bt is also perishable. Most formulations are less effective after a few years of storage.

Bt exits naturally in most soils. Different strains of Bt occur that produce protein crystals toxic to certain insects. The strain for most caterpillars is B.t. var. kurstaki. Commercially prepared Bt spray or powder has no effect on adult butterflies or moths. Remember, however, that not all caterpillars are pests.

Strains of Bt are developed for a few other pests. Some leaf-feeding beetles (including Colorado potato beetle) are susceptible to Bt. tenebrionis , for example.

Because Bt is a near-perfect insecticide there is danger of overuse. Any overused insecticide will gradually become less effective as insects evolve defences to it. Some insect pests, such as the diamondback moth and Indian meal moth, were once susceptible and are now at least partially immune to Bt.

Diatomaceous Earth (DE)

This is a powder-like dust made of the silicate skeletons of tiny marine creatures called diatoms. Millions of years ago as they died, their skeletons gradually accumulated into deep layers that are mined today from deposits where oceans or large lakes once covered the land. DE acts like ground glass, cutting into the waxy coat of insects and causing them to dry out and die. It is not toxic if eaten, but is irritating if inhaled.

Dust DE onto leaves and stems to control pests such as aphids, Colorado potato beetle, immature forms of squash bug, Mexican bean beetle or whitefly. Or spread it as a barrier to slugs and snails. It works best in dry situations. It is not selective and kills spiders and beneficials as well as pests. Don't overuse it. DE is available in two different forms. One form is used primarily in swimming pool filters. It is not an effective insecticide and is dangerous to inhale (it can cause a lung disease called silicosis). In your garden, use only the natural grade of DE. Still, it is wise to wear goggles and a dust mask during application.

Dust DE onto leaves and stems to control pests such as aphids, Colorado potato beetle, immature forms of squash bug, Mexican bean beetle or whitefly. Or spread it as a barrier to slugs and snails. It works best in dry situations. It is not selective and kills spiders and beneficials as well as pests. Don't overuse it.

Horticultural Oils

These are most often highly refined extracts of crude oil. (Some vegetable oils, such as cottonseed and soybean oil, are also sometimes used.) They are increasingly recommended for vegetable garden pest control because they present few risks to either gardeners or desirable species and integrate well with natural biological controls. Also, oils dissipate quickly through evaporation leaving little residue.

Oils kill insects by plugging the pores through which they breathe. Oils can damage plants if applied at excessive rates or on particularly hot (above 100*) or cold (below 40*) days.

Spray oils in vegetable gardens to kill aphids, leafhoppers, spider mites and whiteflies. A few drops of oil in the ear tips of corn controls corn earworm.

Insecticidal Soaps

These are specific fatty acids that have been found by experiment to be toxic to pests, primarily soft-bodied insects such as aphids, mealybugs, spider mites and whiteflies. Surprisingly, adult Japanese beetles are also susceptible. Most nontarget insects are unaffected, and toxicity to animals is non-existent. Soap insecticides act fast and leave no residue. You can use them on vegetables up to the moment of harvest.

Advantages of soaps to home gardeners include safety to both the applicator and non-target insects. They are selective, so easily incorporated with other, natural biological controls. Some plants, such as peas, are readily burned by soaps, and their effectiveness is greatly reduced if mixed with hard water.

Don't use liquid dishwashing detergents or hand soaps. Though many will kill insects, they might hurt the plants too. Some fatty acids are toxic to plants, the reason a soap-based weed killer is now available. Dish soap manufacturers change the oils used in their formulations regularly, based on cost and availability, so the brand that worked fine for your neighbour last year might severely stunt your Brussels sprouts this year.

Malathion

This is a garden insecticide of last resort. Though some consider it relatively safe, many home gardeners avoid using this insecticide simply because it is oily and smells bad. It is, however, widely available and often recommended for control of vegetable garden pests. It is toxic to many kinds of insects, but is less toxic to mammals than either rotenone or Sevin. Though destructive to many beneficial insects, it is less damaging than Sevin.

Neem

This is an extract derived from the crushed seeds of the tropical neem tree (Azadirachta indica). Though intensely studied for many years now, it is still a new botanical insecticide. The primary active ingredient is the compound azadirachtin, although the oils and other ingredients also have some insecticidal effect. A concentrated formulation called Azatin (3% azadirachtin), is available from AgriDyone Technologies in Salt Lake City, UT; (800) 657-3090. Last year, the Environmental Protection Agency stipulated that Azatin was exempt from food crop tolerances because it is considered non-toxic.

Neem works both as an insecticide and as an antifeedant. It kills insects in the juvenile stage by thwarting their development, and is most effective against aphids, thrips and whiteflies. There is no quick "knock-down" with neem, but a week or so after application, you'll notice a steady decline in the number of pests. It is not effective against adult insects (though it may interfere with egg production), and has little impact on beneficial insects. As an antifeedant, neem is effective against Japanese beetles. Apply neem before the beetles appear and reapply after rainfall. Once beetle numbers build up on the plant, neem no longer discourages them.

Neem sprays degrade very quickly in water. Mix only the amount you need and apply all of it immediately. On the plant neem retains its activity against juvenile insects pests for about one week.

Pyrethrins

Derived from the painted daisy, Chrysanthemum cinerariifolium, pyrethrins are considered one of the most important natural insecticides.

When you must use a broad spectrum insecticide in the vegetable garden or lose the crop, this is one of the best choices. Of low toxicity to mammals, they kill insects quickly. In sunlight they break down and are non-toxic within a day or less. For best results apply it in the late afternoon or evening. Use pyrethrins for the hard-to-kill pests such as beetles, squash bugs, and tarnished plant bugs.

The terminology can be confusing. Pyrethrum, discovered around 1800 in the Transcaucasian region of Asia, is the ground-up flowers of the daisy. Pyrethrins (most always plural) are the insecticidal components of the flowers. Pyrethroids , such as cypermethrin, permethrin and resmethrin, are synthetic compounds that resemble pyrethrins. They are more toxic and more persistent than pyrethrins so much more toxic to beneficials. Though increasingly popular in commercially, home gardeners should avoid using them.

Often, pure pyrethrins only stuns insects. This is why they are often combined with a synergist, piperonyl butoxide, chemicals that enhance the effectiveness of the active ingredients, thus enabling formulations with less pyrethrin to kill insects.

Rotenone

This botanical insecticide is extracted from the root of many related tropical legumes, primariely Derris and Lonchocarpus. Though regularly used by organic gardeners, its toxicity to people is greater than common insecticides such as malathion and Sevin, and its residual toxicity lasts up to one week. However, like pyrethrins it does quickly break down in sunlight.

Apply rotenone in the early evening when bees are not active. Like Bt, it is more effective in a pH-balanced solution. Use a buffer solution if your water is strongly alkaline. Rotenone is a broad-spectrum insecticide and should be reserved as a last resort against the hardest to control pests such as cabbageworms, Colorado potato beetles, flea beetles, fruit worms, Japanese beetles, loopers, Mexican bean beetles and weevils.

Sabadilla

This botanical insecticide is made by grinding the seeds of the sabadilla lily (Schoenocaulon officinale) into a fine powder. It's effective against a range of true bugs (Hemiptera) as both a contact and stomach poison. Highly toxic to bees, it is generally considered gentler than rotenone. Sabadilla is highly irritating to mucous membranes of mammals. Like most other botanical insecticides, sabadilla breaks down very quickly in sunlight, but remains potent for many years if stored in a dry, dark place. And like other toxic sprays, use it infrequently, as a last step in your vegetable garden IPM program.

Sevin (Carbaryl)

One of the most widely used insecticides in home vegetable gardens, it is controversial because it kills most kinds of insects. (The Adios product is one way it is used selectively.) Sevin is particularly deadly to honeybees and wasps, several of which are parasites of other garden pests. Because it kills beneficial insects so effectively, Sevin often causes outbreaks of so-called "secondary pests," such as spider mites. It is also toxic to earthworms. Sevin is a sledgehammer of a remedy. Use it sparingly, and only after pyrethrum or perhaps rotenone have failed to be effective. Aside from its effectiveness, Sevin's virtue is low toxicity to mammals and birds. It is less toxic than the botanical insecticides rotenone, and it is less toxic to birds than to mammals. In a garden, however, Sevin requires two weeks or more to degrade, so it poses a significant hazard to a wide range of beneficial insects. Consequently, use it as a last resort, if at all, in your garden IPM program.

Pesticide

Pesticides are substances meant for attracting, seducing, destroying, or mitigating any pest. They are a class of biocide. The most common use of pesticides is as plant protection products (also known as crop protection products), which in general protect plants from damaging influences such as weeds, plant diseases or insects. This use of pesticides is so common that the term *pesticide* is often treated as synonymous with *plant protection product*, although it is in fact a broader term, as pesticides are also used for non-agricultural purposes. The term pesticide includes all of the following: herbicide, insecticide, insect growth regulator, nematicide, termiticide, molluscicide, piscicide, avicide, rodenticide, predacide, bactericide, insect repellent, animal repellent, antimicrobial, fungicide, disinfectant (antimicrobial), and sanitizer.

In general, a pesticide is a chemical or biological agent (such as a virus, bacterium, antimicrobial, or disinfectant) that deters, incapacitates, kills, or otherwise discourages pests. Target pests can include insects, plant pathogens, weeds, mollusks, birds, mammals, fish, nematodes (roundworms), and microbes that destroy property, cause nuisance, or spread disease, or are disease vectors. Although pesticides have benefits, some also have drawbacks, such as potential toxicity to humans and other desired species. According to the Stockholm Convention on Persistent Organic Pollutants, 9 of the 12 most dangerous and persistent organic chemicals are pesticides.

The Food and Agriculture Organisation (FAO) has defined *pesticide* as: *any substance or mixture of substances intended for preventing, destroying, or controlling any pest, including vectors of human or animal disease, unwanted species of plants or animals, causing harm during or otherwise interfering with the production, processing, storage, transport, or marketing of food, agricultural commodities, wood and wood products or animal feedstuffs, or substances that may be administered to animals for the control of insects, arachnids, or other pests in or on their bodies. The term includes substances intended for use as a plant growth regulator, defoliant, desiccant, or agent for thinning fruit or preventing the premature fall of fruit. Also used as substances applied to crops either before or after harvest to protect the commodity from deterioration during storage and transport.*

Pesticides can be classified by target organism (e.g., herbicides, insecticides, fungicides, rodenticides, and pediculicides), chemical structure (e.g., organic, inorganic, synthetic, or biological (biopesticide), although the distinction can sometimes blur), and physical state (e.g. gaseous (fumigant)). Biopesticides include microbial pesticides and biochemical pesticides. Plant-derived pesticides, or "botanicals", have been developing quickly. These include the pyrethroids, rotenoids, nicotinoids, and a fourth group that includes strychnine and scilliroside.

Many pesticides can be grouped into chemical families. Prominent insecticide families include organochlorines, organophosphates, and carbamates. Organochlorine hydrocarbons (e.g., DDT) could be separated into dichlorodiphenylethanes, cyclodiene compounds, and other related compounds. They operate by disrupting the sodium/potassium balance of the nerve fibre, forcing the nerve to transmit continuously. Their toxicities vary greatly, but they have been phased out because of their persistence and potential to bioaccumulate. Organophosphate and carbamates largely replaced organochlorines. Both operate through inhibiting the enzyme acetylcholinesterase, allowing acetylcholine to transfer nerve impulses indefinitely and causing a variety of symptoms such as weakness or paralysis. Organophosphates are quite toxic to vertebrates, and have in some cases been replaced by less toxic carbamates. Thiocarbamate and dithiocarbamates are subclasses of carbamates. Prominent families of herbicides include phenoxy and benzoic acid herbicides (e.g. 2,4-D), triazines (e.g., atrazine), ureas (e.g., diuron), and Chloroacetanilides (e.g., alachlor). Phenoxy compounds tend to selectively kill broad-leaf weeds rather than grasses. The phenoxy and benzoic acid herbicides function similar to plant growth hormones, and grow cells without normal cell division, crushing the plant's nutrient transport system. Triazines interfere with photosynthesis. Many commonly used pesticides are not included in these families, including glyphosate.

Pesticides can be classified based upon their biological mechanism function or application method. Most pesticides work by poisoning pests. A systemic pesticide moves inside a plant following absorption by the plant. With insecticides and most fungicides, this movement is usually upward (through the xylem) and outward. Increased efficiency may be a result. Systemic insecticides, which poison pollen and nectar in the flowers, may kill bees and other needed pollinators.

In 2009, the development of a new class of fungicides called paldoxins was announced. These work by taking advantage of natural defence chemicals released by plants called phytoalexins, which fungi then detoxify using enzymes. The paldoxins inhibit the fungi's detoxification enzymes. They are believed to be safer and greener.

Types

Pesticides are often referred to according to the type of pest they control. Pesticides can also be considered as either biodegradable pesticides, which will be broken down by microbes and other living beings into harmless compounds, or persistent pesticides, which may take months or years before they are broken down: it was the persistence of DDT, for example, which led to its accumulation in the food chain and its killing of birds of prey at the top of the food chain. Another way to think about pesticides is to consider those that are chemical pesticides or are derived from a common source or production method.

Some examples of chemically-related pesticides are:

Organophosphate Pesticides: Organophosphates affect the nervous system by disrupting the enzyme that regulates acetylcholine, a neurotransmitter. Most organophosphates are insecticides. They were developed during the early 19th century, but their effects on insects, which are similar to their effects on humans, were discovered in 1932. Some are very poisonous (they were used in World War II as nerve agents). However, they usually are not persistent in the environment.

Carbamate Pesticides: Carbamate pesticides affect the nervous system by disrupting an enzyme that regulates acetylcholine, a neurotransmitter. The enzyme effects are usually reversible. There are several subgroups within the carbamates.

Organochlorine Insecticides: They were commonly used in the past, but many have been removed from the market due to their health and environmental effects and their persistence (e.g., DDT and chlordane).

Pyrethroid Pesticides: They were developed as a synthetic version of the naturally occurring pesticide pyrethrin, which is found in chrysanthemums. They have been modified to increase their stability in the environment. Some synthetic pyrethroids are toxic to the nervous system.

Sulfonylurea Herbicides: Includes nicosulfuron, triflusulfuron methyl, and chlorsulfuron broad-spectrum herbicides that kill plants by inhibiting the enzyme acetolactate synthase. In the 1960s, more than 1 kg/ha (0.89 lb/acre) crop protection chemical was typically applied, while sulfonylureates allow as little as 1% as much material to achieve the same effect.

Biopesticides: Biopesticides are certain types of pesticides derived from such natural materials as animals, plants, bacteria, and certain minerals. For example, canola oil and baking soda have pesticidal applications and are considered biopesticides. At the end of 2001, there were approximately 195 registered biopesticide active ingredients and 780 products. Biopesticides fall into three major classes:

- Microbial pesticides consist of a microorganism e.g., a bacterium, fungus, virus, or protozoan as the active ingredient. Microbial pesticides can control many different kinds of pests, although each separate active ingredient is relatively specific for its target pest. For example, there are fungi that control certain weeds, and other fungi that kill specific insects.

The most widely used microbial pesticides are subspecies and strains of Bacillus thuringiensis, or Bt. Each strain of this bacterium produces a different mix of proteins, and specifically kills one or a few related species of insect larvae. While some Bt's control moth larvae found on plants, other Bt's are specific for larvae of flies and mosquitoes. The target insect species are determined by whether the particular Bt produces a protein that can bind to a larval gut receptor, thereby causing the insect larvae to starve.

- Plant-Incorporated-Protectants (PIPs) are pesticidal substances that plants produce from genetic material that has been added to the plant. For example, scientists can take the gene for the Bt pesticidal protein, and introduce the gene into the plant's own genetic material. Then the plant, instead of the Bt bacterium, manufactures the substance that destroys the pest. The protein and its genetic material, but not the plant itself, are regulated by EPA.
- Biochemical pesticides are naturally occurring substances that control pests by non-toxic mechanisms. Conventional pesticides,

by contrast, are, in general, synthetic materials that directly kill or inactivate the pest. Biochemical pesticides include substances, such as insect sex pheromones, that interfere with mating, as well as various scented plant extracts that attract insect pests to traps.

Because it is sometimes difficult to determine whether a substance meets the criteria for classification as a biochemical pesticide, EPA has established a special committee to make such decisions.

Classified by Type of Pest

Pesticides that are related to the type of pests are:

Type	*Action*
Algicides	Control algae in lakes, canals, swimming pools, water tanks, and other sites
Antifouling agents	Kill or repel organisms that attach to underwater surfaces, such as boat bottoms
Antimicrobials	Kill microorganisms (such as bacteria and viruses)
Attractants	Attract pests (for example, to lure an insect or rodent to a trap). (However, food is not considered a pesticide when used as an attractant.)
Biopesticides	Biopesticides are certain types of pesticides derived from such natural materials as animals, plants, bacteria, and certain minerals
Biocides	Kill microorganisms Disinfectants and sanitizers Kill or inactivate disease-producing microorganisms on inanimate objects
Fungicides	Kill fungi (including blights, mildews, molds, and rusts)
Fumigants	Produce gas or vapor intended to destroy pests in buildings or soil
Herbicides	Kill weeds and other plants that grow where they are not wanted
Insecticides	Kill insects and other arthropods
Miticides	Kill mites that feed on plants and animals
Microbial pesticides	Microorganisms that kill, inhibit, or out compete pests, including insects or other microorganisms
Molluscicides	Kill snails and slugs
Nematicides	Kill nematodes (microscopic, worm-like organisms that feed on plant roots)
Ovicides	Kill eggs of insects and mites

Contd...

Type	***Action***
Pheromones	Biochemicals used to disrupt the mating behaviour of insects
Repellents	Repel pests, including insects (such as mosquitoes) and birds
Rodenticides	Control mice and other rodents

Further Types of Pesticides

The term pesticide also include these substances:

Defoliants: Cause leaves or other foliage to drop from a plant, usually to facilitate harvest.

Desiccants: Promote drying of living tissues, such as unwanted plant tops.

Insect Growth Regulators: Disrupt the molting, maturity from pupal stage to adult, or other life processes of insects.

Plant Growth Regulators: Substances (excluding fertilizers or other plant nutrients) that alter the expected growth, flowering, or reproduction rate of plants.

Uses

Pesticides are used to control organisms that are considered to be harmful. For example, they are used to kill mosquitoes that can transmit potentially deadly diseases like West Nile virus, yellow fever, and malaria. They can also kill bees, wasps or ants that can cause allergic reactions. Insecticides can protect animals from illnesses that can be caused by parasites such as fleas. Pesticides can prevent sickness in humans that could be caused by moldy food or diseased produce. Herbicides can be used to clear roadside weeds, trees and brush. They can also kill invasive weeds that may cause environmental damage. Herbicides are commonly applied in ponds and lakes to control algae and plants such as water grasses that can interfere with activities like swimming and fishing and cause the water to look or smell unpleasant. Uncontrolled pests such as termites and mould can damage structures such as houses.

Pesticides are used in grocery stores and food storage facilities to manage rodents and insects that infest food such as grain. Each use of a pesticide carries some associated risk. Proper pesticide use decreases these associated risks to a level deemed acceptable by pesticide regulatory agencies such as the United States Environmental Protection Agency (EPA) and the Pest Management Regulatory Agency (PMRA) of Canada.

Pesticides can save farmers' money by preventing crop losses to insects and other pests; in the U.S., farmers get an estimated fourfold return on money they spend on pesticides. One study found that not using pesticides reduced crop yields by about 10%. Another study, conducted in 1999, found that a ban on pesticides in the United States may result in a rise of food prices, loss of jobs, and an increase in world hunger.

DDT, sprayed on the walls of houses, is an organochloride that has been used to fight malaria since the 1950s. Recent policy statements by the World Health Organisation have given stronger support to this approach. Dr. Arata Kochi, WHO's malaria chief, said, "One of the best tools we have against malaria is indoor residual house spraying. Of the dozen insecticides WHO has approved as safe for house spraying, the most effective is DDT." However a later October 2007 study linked breast cancer from exposure to DDT prior to puberty. Poisoning may also occur due to use of DDT and other chlorinated hydrocarbons by entering the human food chain when animal tissues are affected. Symptoms include nervous excitement, tremors, convulsions, or death. Scientists estimate that DDT and other chemicals in the organophosphate class of pesticides have saved 7 million human lives since 1945 by preventing the transmission of diseases such as malaria, bubonic plague, sleeping sickness, and typhus. However, DDT use is not always effective, as resistance to DDT was identified in Africa as early as 1955, and by 1972 nineteen species of mosquito worldwide were resistant to DDT. A study for the World Health Organisation in 2000 from Vietnam established that non-DDT malaria controls were significantly more effective than DDT use. The ecological effect of DDT on organisms is an example of bioaccumulation.

Manufacturing Process of Pesticides

Quantity and Variety

In 2006 and 2007, the world used approximately 5.2 billion pounds of pesticides, with herbicides constituting the biggest part of the world pesticide use at 40%, followed by insecticides (17%) and fungicides (10%). In 2006 and 2007 the U.S. used approximately 1.1 billion pounds of pesticides, accounting for 22% of the world total, including 857 million pounds of conventional pesticides, which are used in the agricultural sector (80% of conventional pesticide use) as well as the industrial, commercial, governmental and home & garden sectors. Pesticides are also found in majority of U.S. households with 78 million out of the 105.5 million households indicating that they use some form of pesticide. As of 2007, there were more than 1,055 active ingredients registered as pesticides, which yield over 20,000 pesticide products that are marketed in the United States.

The US used some 2.2 pounds per hactare of arable land compared with 10.3 in China, 3.0 in the UK, .2 in Cameroun, 13.1 in Japan and 5.6 in Italy. Insecticide use in the US has declined by more than half since 1980, (.6%/yr) mostly duc to the near phase-out of organophosphates. In corn fields, the decline was even steeper, due to the switchover to transgenic Bt corn.

For the global market of crop protection products, market analysts forecast revenues of over 52 billion US$ in 2019.

Costs

On the cost side of pesticide use there can be costs to the environment, costs to human health, as well as costs of the development and research of new pesticides.

Health Effects

Figure: *A sign warning about potential pesticide exposure.*

Pesticides may cause acute and delayed health effects in workers who are exposed. Pesticide exposure can cause a variety of adverse health effects, ranging from simple irritation of the skin and eyes to more severe effects such as affecting the nervous system, mimicking hormones causing reproductive problems, and also causing cancer. A 2007 systematic review found that “most studies on non-Hodgkin lymphoma and leukemia showed positive associations with pesticide exposure” and thus concluded that cosmetic use of pesticides should be decreased. Strong evidence also exists for other negative outcomes from pesticide exposure including neurological, birth defects, fetal death, and neurodevelopmental disorder.

The American Medical Association recommends limiting exposure to pesticides and using safer alternatives: “Particular uncertainty exists regarding the long-term effects of low-dose pesticide exposures.”

Even consuming organic food, these chemicals can have a negative health impact. Though the U.S. Environmental Protection Agency (EPA) limits the amount of each pesticide that may be present on a food item, there exists no limit to the number of different chemicals that can be used. This leads to a possibly even more dangerous effect known as a "chemical cocktail". The chemicals may form one of many dangerous interactions and have an unmeasured synergistic effect.

The World Health Organisation and the UN Environment Programme estimate that each year, 3 million workers in agriculture in the developing world experience severe poisoning from pesticides, about 18,000 of whom die. According to one study, as many as 25 million workers in developing countries may suffer mild pesticide poisoning yearly. One study found pesticide self-poisoning the method of choice in one third of suicides worldwide, and recommended, among other things, more restrictions on the types of pesticides that are most harmful to humans.

A 2007 study by the California Department of Public Health found that women in the first eight weeks of pregnancy who live near farm fields sprayed with the organochlorine pesticides dicofol and endosulfan are several times more likely to give birth to children with autism.

Environmental Effect

Pesticide use raises a number of environmental concerns. Over 98% of sprayed insecticides and 95% of herbicides reach a destination other than their target species, including non-target species, air, water and soil. Pesticide drift occurs when pesticides suspended in the air as particles are carried by wind to other areas, potentially contaminating them. Pesticides are one of the causes of water pollution, and some pesticides are persistent organic pollutants and contribute to soil contamination.

In addition, pesticide use reduces biodiversity, reduces nitrogen fixation, contributes to pollinator decline, destroys habitat (especially for birds), and threatens endangered species. Pests can develop a resistance to the pesticide (pesticide resistance), necessitating a new pesticide. Alternatively a greater dose of the pesticide can be used to counteract the resistance, although this will cause a worsening of the ambient pollution problem.

Since chlorinated hydrocarbon pesticides dissolve in fats and are not excreted, organisms tend to retain them almost indefinitely. Biological magnification is the process whereby these chlorinated hydrocarbons (pesticides) are more concentrated at each level of the food chain. Among marine animals, pesticide concentrations are higher in carnivorous fishes, and even more so in the fish-eating birds and mammals at the top of the ecological pyramid.

Global distillation is the process whereby pesticides are transported from warmer to colder regions of the Earth, in particular the Poles and mountain tops. Pesticides that evaporate into the atmosphere at relatively high temperature can be carried considerable distances (thousands of kilometres) by the wind to an area of lower temperature, where they condense and are carried back to the ground in rain or snow.

In order to reduce negative impacts, it is desirable that pesticides be degradable or at least quickly deactivated in the environment. Such loss of activity or toxicity of pesticides is due to both innate chemical properties of the compounds and environmental processes or conditions. For example, the presence of halogens within a chemical structure often slows down degradation in an aerobic environment. Adsorption to soil may retard pesticide movement, but also may reduce bioavailability to microbial degraders.

Economics

Harm	*Annual US Cost*
Public Health	\$1.1 billion
Pesticide Resistance in Pest	\$1.5 billion
Crop Losses Caused by Pesticides	\$1.4 billion
Bird Losses due to Pesticides	\$2.2 billion
Groundwater Contamination	\$2.0 billion
Other Costs	\$1.4 billion
Total Costs	\$9.6 billion

Human health and environmental cost from pesticides in the United States is estimated at \$9.6 billion:

Additional costs include the registration process and the cost of purchasing pesticides. The registration process can take several years to complete (there are 70 different types of field test) and can cost \$50–70 million for a single pesticide. Annually the United States spends \$10 billion on pesticides.

Benefits

There are two levels of benefits for pesticide use, primary and secondary. Primary benefits are direct gains from the use of pesticides and secondary benefits are effects that are more long-term.

Primary Benefits

Controlling pests and plant disease vectors

- Improved crop/livestock yields
- Improved crop/livestock quality
- Invasive species controlled

Controlling human / livestock disease vectors and nuisance organisms

- Human lives saved and suffering reduced
- Animal lives saved and suffering reduced
- Diseases contained geographically

Controlling organisms that harm other human activities and structures

- Drivers view unobstructed
- Tree/brush/leaf hazards prevented
- Wooden structures protected

Monetary

Every dollar ($1) that is spent on pesticides for crops yields four dollars ($4) in crops saved. This means based that, on the amount of money spent per year on pesticides, $10 billion, there is an additional $40 billion savings in crop that would be lost due to damage by insects and weeds. In general, farmers benefit from having an increase in crop yield and from being able to grow a variety of crops throughout the year. Consumers of agricultural products also benefit from being able to afford the vast quantities of produce available year-round. The general public also benefits from the use of pesticides for the control of insect-borne diseases and illnesses, such as malaria. The use of pesticides creates a large job market, which provides jobs for all of the people working within the industry.

Alternatives

Alternatives to pesticides are available and include methods of cultivation, use of biological pest controls (such as pheromones and microbial pesticides), genetic engineering, and methods of interfering with insect breeding. Application of composted yard waste has also been used as a way of controlling pests. These methods are becoming increasingly popular and often are safer than traditional chemical pesticides. In addition, EPA is registering reduced-risk conventional pesticides in increasing numbers.

Cultivation practices include polyculture (growing multiple types of plants), crop rotation, planting crops in areas where the pests that damage them do not live, timing planting according to when pests will be least problematic, and use of trap crops that attract pests away from the real crop. In the U.S., farmers have had success controlling insects by spraying with hot water at a cost that is about the same as pesticide spraying.

Release of other organisms that fight the pest is another example of an alternative to pesticide use. These organisms can include natural predators or parasites of the pests. Biological pesticides based on entomopathogenic fungi, bacteria and viruses cause disease in the pest species can also be used.

Interfering with insects' reproduction can be accomplished by sterilizing males of the target species and releasing them, so that they mate with females but do not produce offspring. This technique was first used on the screwworm fly in 1958 and has since been used with the medfly, the tsetse fly, and the gypsy moth. However, this can be a costly, time consuming approach that only works on some types of insects.

Another alternative to pesticides is the thermal treatment of soil through steam. Soil steaming kills pest and increases soil health.

In India, traditional pest control methods include using Panchakavya, the "mixture of five products." The method has recently experienced a resurgence in popularity due in part to use by the organic farming community.

Agroecology emphasize nutrient recycling, use of locally available and renewable resources, adaptation to local conditions, utilisation of microenvironments, reliance on indigenous knowledge and yield maximization while maintaining soil productivity. Agroecology also emphasizes empowering people and local communities to contribute to development, and encouraging "multi-directional" communications rather than the conventional "top-down" method.

Push Pull Strategy

The term "push-pull" was established in 1987 as an approach for integrated pest management (IPM). This strategy uses a mixture of behaviour-modifying stimuli to manipulate the distribution and abundance of insects. "Push" means the insects are repelled or deterred away from whatever resource that is being protected. "Pull" means that certain stimuli (semiochemical stimuli, pheromones, food additives, visual stimuli, genetically altered plants, etc.) are used to attract pests to trap crops where they will be killed. There are numerous different components involved in order to implement a Push-Pull Strategy in IPM.

Many case studies testing the effectiveness of the push-pull approach have been done across the world. The most successful push-pull strategy was developed in Africa for subsistence farming. Another successful case study was performed on the control of *Helicoverpa* in cotton crops in Australia. In Europe, the Middle East, and the United States, push-pull strategies were successfully used in the controlling of *Sitona lineatus* in bean fields.

Some advantages of using the push-pull method are less use of chemical or biological materials and better protection against insect habituation to this control method. Some disadvantages of the push-pull strategy is that if there is a lack of appropriate knowledge of behavioural and chemical ecology of the host-pest interactions then this method becomes unreliable. Furthermore, because the push-pull method is not a very popular method of IPM operational and registration costs are higher.

Effectiveness

Some evidence shows that alternatives to pesticides can be equally effective as the use of chemicals. For example, Sweden has halved its use of pesticides with hardly any reduction in crops. In Indonesia, farmers have reduced pesticide use on rice fields by 65% and experienced a 15% crop increase. A study of Maize fields in northern Florida found that the application of composted yard waste with high carbon to nitrogen ratio to agricultural fields was highly effective at reducing the population of plant-parasitic nematodes and increasing crop yield, with yield increases ranging from 10% to 212%; the observed effects were long-term, often not appearing until the third season of the study.

However, pesticide resistance is increasing. In the 1940s, U.S. farmers lost only 7% of their crops to pests. Since the 1980s, loss has increased to 13%, even though more pesticides are being used. Between 500 and 1,000 insect and weed species have developed pesticide resistance since 1945.

Regulation

International: In most countries, pesticides must be approved for sale and use by a government agency.

In Europe, recent EU legislation has been approved banning the use of highly toxic pesticides including those that are carcinogenic, mutagenic or toxic to reproduction, those that are endocrine-disrupting, and those that are persistent, bioaccumulative and toxic (PBT) or very persistent and very bioaccumulative (vPvB). Measures were approved to improve the general safety of pesticides across all EU member states.

Though pesticide regulations differ from country to country, pesticides, and products on which they were used are traded across international borders. To deal with inconsistencies in regulations among countries, delegates to a conference of the United Nations Food and Agriculture Organisation adopted an International Code of Conduct on the Distribution and Use of Pesticides in 1985 to create voluntary standards of pesticide regulation for different countries.

The Code was updated in 1998 and 2002. The FAO claims that the code has raised awareness about pesticide hazards and decreased the number of countries without restrictions on pesticide use.

Three other efforts to improve regulation of international pesticide trade are the United Nations London Guidelines for the Exchange of Information on Chemicals in International Trade and the United Nations Codex Alimentarius Commission . The former seeks to implement procedures for ensuring that prior informed consent exists between countries buying and selling pesticides, while the latter seeks to create uniform standards for maximum levels of pesticide residues among participating countries. Both initiatives operate on a voluntary basis.

Pesticide safety education and pesticide applicator regulation are designed to protect the public from pesticide misuse, but do not eliminate all misuse. Reducing the use of pesticides and choosing less toxic pesticides may reduce risks placed on society and the environment from pesticide use. Integrated pest management, the use of multiple approaches to control pests, is becoming widespread and has been used with success in countries such as Indonesia, China, Bangladesh, the U.S., Australia, and Mexico. IPM attempts to recognise the more widespread impacts of an action on an ecosystem, so that natural balances are not upset. New pesticides are being developed, including biological and botanical derivatives and alternatives that are thought to reduce health and environmental risks. In addition, applicators are being encouraged to consider alternative controls and adopt methods that reduce the use of chemical pesticides.

Pesticides can be created that are targeted to a specific pest's lifecycle, which can be environmentally more friendly. For example, potato cyst nematodes emerge from their protective cysts in response to a chemical excreted by potatoes; they feed on the potatoes and damage the crop.

A similar chemical can be applied to fields early, before the potatoes are planted, causing the nematodes to emerge early and starve in the absence of potatoes.

United States

Figure: *Preparation for an application of hazardous herbicide in USA.*

In the United States, the Environmental Protection Agency (EPA) is responsible for regulating pesticides under the Federal Insecticide, Fungicide, and Rodenticide Act (FIFRA) and the Food Quality Protection Act (FQPA). Studies must be conducted to establish the conditions in which the material is safe to use and the effectiveness against the intended pest(s). The EPA regulates pesticides to ensure that these products do not pose adverse effects to humans or the environment. Pesticides produced before November 1984 continue to be reassessed in order to meet the current scientific and regulatory standards. All registered pesticides are reviewed every 15 years to ensure they meet the proper standards. During the registration process, a label is created. The label contains directions for proper use of the material in addition to safety restrictions. Based on acute toxicity, pesticides are assigned to a Toxicity Class.

Some pesticides are considered too hazardous for sale to the general public and are designated restricted use pesticides. Only certified applicators, who have passed an exam, may purchase or supervise the application of restricted use pesticides. Records of sales and use are required to be maintained and may be audited by government agencies charged with the enforcement of pesticide regulations. These records must be made available to employees and state or territorial environmental regulatory agencies.

The EPA regulates pesticides under two main acts, both of which amended by the Food Quality Protection Act of 1996. In addition to the EPA, the United States Department of Agriculture (USDA) and the United States Food and Drug Administration (FDA) set standards for the level of pesticide residue that is allowed on or in crops. The EPA looks at what the potential human health and environmental effects might be associated with the use of the pesticide. In addition, the U.S. EPA uses the National Research Council's four-step process for human health risk assessment:

(1) Hazard Identification,

(2) Dose-Response Assessment,

(3) Exposure Assessment, and

(4) Risk Characterization.

Recently Kaua'i County (Hawai'i) passed Bill No. 2491 to add an article to Chapter 22 of the county's code relating to pesticides and GMOs. The bill strengthens protections of local communities in Kaua'i where many large pesticide companies test their products.

History

Since before 2000 BC, humans have utilised pesticides to protect their crops. The first known pesticide was elemental sulfur dusting used in ancient Sumer about 4,500 years ago in ancient Mesopotamia. The Rig Veda, which is about 4,000 years old, mentions the use of poisonous plants for pest control. By the 15th century, toxic chemicals such as arsenic, mercury, and lead were being applied to crops to kill pests. In the 17th century, nicotine sulfate was extracted from tobacco leaves for use as an insecticide.

The 19th century saw the introduction of two more natural pesticides, pyrethrum, which is derived from chrysanthemums, and rotenone, which is derived from the roots of tropical vegetables. Until the 1950s, arsenic-based pesticides were dominant. Paul Müller discovered that DDT was a very effective insecticide. Organochlorines such as DDT were dominant, but they were replaced in the U.S. by organophosphates and carbamates by 1975. Since then, pyrethrin compounds have become the dominant insecticide. Herbicides became common in the 1960s, led by "triazine and other nitrogen-based compounds, carboxylic acids such as 2,4-dichlorophenoxyacetic acid, and glyphosate".

The first legislation providing federal authority for regulating pesticides was enacted in 1910; however, decades later during the 1940s manufacturers began to produce large amounts of synthetic pesticides and their use became widespread. Some sources consider the 1940s and 1950s to have been the start of the "pesticide era." Although the U.S. Environmental Protection Agency was established in 1970 and amendments to the pesticide law in 1972, pesticide use has increased 50-fold since 1950 and 2.3 million tonnes (2.5 million short tons) of industrial pesticides are now used each year. Seventy-five percent of all pesticides in the world are used in developed countries, but use in developing countries is increasing. In 2001 the EPA stopped reporting yearly pesticide use statistics. A study of USA pesticide use trends through 1997 was published in 2003 by the National Science Foundation's Centre for Integrated Pest Management.

In the 1960s, it was discovered that DDT was preventing many fish-eating birds from reproducing, which was a serious threat to biodiversity. Rachel Carson wrote the best-selling book Silent Spring about biological magnification. The agricultural use of DDT is now banned under the Stockholm Convention on Persistent Organic Pollutants, but it is still used in some developing nations to prevent malaria and other tropical diseases by spraying on interior walls to kill or repel mosquitoes.

Pest Control

Pest control refers to the regulation or management of a species defined as a pest, usually because it is perceived to be detrimental to a person's health, the ecology or the economy.

Practitioners of pest control are referred to as Exterminators.

Pest control is at least as old as agriculture, as there has always been a need to keep crops free from pests. In order to maximize food production, it is advantageous to protect crops from competing species of plants, as well as from herbivores competing with humans.

The conventional approach was probably the first to be employed, since it is comparatively easy to destroy weeds by burning them or plowing them under, and to kill larger competing herbivores, such as crows and other birds eating seeds. Techniques such as crop rotation, companion planting (also known as intercropping or mixed cropping), and the selective breeding of pest-resistant cultivars have a long history.

In the UK, following concern about animal welfare, humane pest control and deterrence is gaining ground through the use of animal psychology rather than destruction. For instance, with the urban Red Fox which territorial behaviour is used against the animal, usually in conjunction with non-injurious chemical repellents. In rural areas of Britain, the use of firearms for pest control is quite common. Airguns are particularly popular for control of small pests such as rats, rabbits and grey squirrels, because of their lower power they can be used in more restrictive spaces such as gardens, where using a firearm would be unsafe.

Chemical pesticides date back 4,500 years, when the Sumerians used sulfur compounds as insecticides. The Rig Veda, which is about 4,000 years old, also mentions the use of poisonous plants for pest control. It was only with the industrialization and mechanization of agriculture in the 18th and 19th century, and the introduction of the insecticides pyrethrum and derris that chemical pest control became widespread. In the 20th century, the discovery of several synthetic insecticides, such as DDT, and herbicides boosted this development. Chemical pest control is still the predominant type of pest control today, although its long-term effects led to a renewed interest in traditional and biological pest control towards the end of the 20th century.

Causes

Many pests have only become a problem because of the direct actions of humans. Modifying these actions can often substantially reduce the pest problem. In the United States, raccoons caused a nuisance by tearing open refuse sacks. Many householders introduced bins with locking lids, which deterred the raccoons from visiting. House flies tend to accumulate wherever there is human activity and is virtually a global phenomenon, especially where food or food waste is exposed. Similarly, seagulls have become pests at many seaside resorts. Tourists would often feed the birds with scraps of fish and chips, and before long, the birds would become dependent on this food source and act aggressively towards humans.

Living organisms evolve and increase their resistance to biological, chemical, physical or any other form of control. Unless the target population is completely exterminated or is rendered incapable of reproduction, the surviving population will inevitably acquire a tolerance of whatever pressures are brought to bear - this results in an evolutionary arms race.

Types of Pest Control

Biological Pest Control

Biological pest control is the control of one through the control and management of natural predators and parasites. For example: mosquitoes are often controlled by putting *Bt Bacillus thuringiensis* ssp. *israelensis*, a bacterium that infects and kills mosquito larvae, in local water sources. The treatment has no known negative consequences on the remaining ecology and is safe for humans to drink. The point of biological pest control, or any natural pest control, is to eliminate a pest with minimal harm to the ecological balance of the environment in its present form.

Mechanical Pest Control

Mechanical pest control is the use of hands-on techniques as well as simple equipment, devices, and natural ingredients that provide a protective barrier between plants and insects. For example: weeds can be controlled by being physically removed from the ground. This is referred to as tillage and is one of the oldest methods of weed control.

Physical Pest Control

Physical pest control is a method of getting rid of insects and small rodents by removing, attacking, or setting up barriers that will prevent further destruction of one's plants.

Elimination of Breeding Grounds

Proper waste management and drainage of still water, eliminates the breeding ground of many pests.

Garbage provides food and shelter for many unwanted organisms, as well as an area where still water might collect and be used as a breeding ground by mosquitoes. Communities that have proper garbage collection and disposal, have far less of a problem with rats, cockroaches, mosquitoes, flies and other pests than those that don't.

Open air sewers are ample breeding ground for various pests as well. By building and maintaining a proper sewer system, this problem is eliminated.

Certain spectrums of LED light can "disrupt insects' breeding.

Poisoned Bait

Poisoned bait is a common method for controlling rat populations, however is not as effective when there are other food sources around, such as garbage. Poisoned meats have been used for centuries for killing off wolves, birds that were seen to threaten crops, and against other creatures. This can be a problem, since a carcass which has been poisoned will kill not only the targeted animal, but also every other animal which feeds on the carcass. Humans have also been killed by coming in contact with poisoned meat, or by eating an animal which had fed on a poisoned carcass. This tool is also used to manage several caterpillars e.g.Spodoptera litura, fruit flies, snails and slugs, crabs etc.

Field Burning

Traditionally, after a sugar cane harvest, the fields are all burned, to kill off any insects or eggs that might be in the fields. Examples of these pests are Rodents.

Hunting

Historically, in some European countries, when stray dogs and cats became too numerous, local populations gathered together to round up all animals that did not appear to have an owner and kill them. In some nations, teams of rat catchers work at chasing rats from the field, and killing them with dogs and simple hand tools. Some communities have in the past employed a bounty system, where a town clerk will pay a set fee for every rat head brought in as proof of a rat killing.

Traps

With the many traps available on the market today you can easily remove mice and rats from homes. You must first know what rodent needs to be removed, you can then decide what type of trap is the best suited to your needs. The snap trap is the most widely used, it utilises a trigger (sometimes shaped like cheese) to hold bait, and kills the rodent by striking it behind the head with a wire rod or jaw. In some instances you may wish to use glue traps also called glue boards. This type of trap requires the mouse or rat to attempt to cross the trap so the glue can hold the rodent. After a catch is made you can euthanize the rodent and dispose of it trap and all, or some glue boards will release the catch when you pour vegetable oil on them, as the oil reacts with the glue to lose its grip. The last type of trap are live catch traps, this type of trap is typically a repeating style so more than one animal can be caught at a time, they can also be released from this trap in a new location if desired.

Pesticides

Spraying pesticides by planes, handheld units, or trucks that carry the spraying equipment, is a common method of pest control. Crop dusters commonly fly over farmland and spray pesticides to kill off pests that would threaten the crops. However, some pesticides may cause cancer and other health problems, as well as harming wildlife.

Space Fumigation

A project that involves a structure be covered or sealed airtight followed by the introduction of a penetrating, deadly gas at a killing concentration a long period of time (24-72hrs.). Although expensive, space fumigation targets all life stages of pests.

Space Treatment

A long term project involving fogging or misting type applicators. Liquid insecticide is dispersed in the atmosphere within a structure. Treatments do not require the evacuation or airtight sealing of a building, allowing most work within the building to continue but at the cost of the penetrating effects. Contact insecticides are generally used, minimizing the long lasting residual effects. On August 10, 1973, the Federal Register printed the definition of Space treatment as defined by the U.S. Environmental Protection Agency (EPA): the dispersal of insecticides into the air by foggers, misters, aerosol devices or vapor dispensers for control of flying insects and exposed crawling insects

Figure: *Residential & commercial building pest control service vehicle, Ypsilanti Township, Michigan*

Sterilization

Laboratory studies conducted with U-5897 (3-chloro-1,2-propanediol) where attempted in the early 1970s although these proved unsuccessful. Research into sterilization bait is ongoing. Another effective method of soil sterilization is soil steaming. Pest is killed through hot steam which is induced into the soil.

Destruction of Infected Plants

Forest services sometimes destroy all the trees in an area where some are infected with insects, if seen as necessary to prevent the insect species from spreading. Farms infested with certain insects, have been burned entirely, to prevent the pest from spreading elsewhere.

Natural Rodent Control

Several wildlife rehabilitation organisations encourage natural form of rodent control through exclusion and predator support and preventing secondary poisoning altogether. The United States Environmental Protection Agency agrees, noting in its Proposed Risk Mitigation Decision for Nine Rodenticides that "without habitat modification to make areas less attractive to commensal rodents, even eradication will not prevent new populations from recolonizing the habitat."

Repellents

- Balsam fir oil from the tree *Abies balsamea* is an EPA approved non-toxic rodent repellent.

- *Acacia polyacantha* subsp. *campylacantha* root emits chemical compounds that repel animals including crocodiles, snakes and rats.

Pesticide Residue in Vegetable Crops

Pesticide residue refers to the pesticides that may remain on or in food after they are applied to food crops. The levels of these residues in foods are often stipulated by regulatory bodies in many countries. Exposure of the general population to these residues most commonly occurs through consumption of treated food sources, or being in close contact to areas treated with pesticides such as farms or lawns around houses. Many of these chemical residues, especially derivatives of chlorinated pesticides, exhibit bioaccumulation which could build up to harmful levels in the body as well as in the environment. Persistent chemicals can be magnified through the food chain and have been detected in products ranging from meat, poultry, and fish, to vegetable oils, nuts, and various fruits and vegetables.

From post-World War II era, chemical pesticides have become the most important form of pest control. There are two categories of pesticides, first-generation pesticides and second-generation pesticide. The first-generation pesticides, which were used prior to 1940, consisted of compounds such as arsenic, mercury, and lead. These were soon abandoned because they were highly toxic and ineffective. The second-generation pesticides were composed of synthetic organic compounds. The growth in these pesticides accelerated in late 1940s after Paul Müller discovered DDT in 1939. The effects of pesticides such as aldrin, dieldrin, endrin, chlordane, parathion, captan and 2,4-D were also found at this time. Those pesticides were widely used due to its effective pest control. However, in 1946, people started to resist to the widespread use of pesticides, especially DDT since it harms non-target plants and animals. People became aware of problems with residues and its potential health risks. In the 1960s, Rachel Carson wrote *Silent Spring* to illustrate a risk of DDT and how it is threatening biodiversity.

Regulations

Each country adopts their own agricultural policies and Maximum Residue Limits (MRL) and Acceptable Daily Intake (ADI). The level of food additive usage varies by country because forms of agriculture are different in regions according to their geographical or climatical factors.

International

Some countries use the International Maximum Residue Limits - Codex Alimentarius to define the residue limits; this was established by Food and Agriculture Organisation of the United Nations (FAO) and World Health Organisation (WHO) in 1963 to develop international food standards, guidelines codes of practices, and recommendation for food safety. Currently the CODEX has 185 Member Countries and 1 member organisation (EU).

The following is the list of maximum residue limits (MRLs) for spices adopted by the commission.

Pesticide	*Group or Sub-Group of Spices*	*MRL (mg/kg)*
Acephate	Entire Group 028	0.2
Azinphos-methyl	Entire Group 028	0.5
Chlorpyrifos	SeedsFruits or berriesRoots or rhizomes	511
Chlorpyrifos-methyl	SeedsFruits or berriesRoots or rhizomes	10.35
Cypermethrin	Fruits or berriesRoots or rhizomes	0.10.2
Diazinon	SeedsFruitsRoots or rhizomes	50.10.5
Dichlorvos	Entire Group 028	0.1
Dicofol	SeedsFruits or berriesRoots or rhizomes	0.050.10.1
Dimethoate	SeedsFruits or berriesRoots or rhizomes	50.50.1
Disulfoton	Entire Group 028	0.05
Endosulfan	SeedsFruits or berriesRoots or rhizomes	150.5
Ethion	SeedsFruits or berriesRoots or rhizomes	350.3
Fenitrothion	SeedsFruits or berriesRoots or rhizomes	710.1
Iprodione	SeedsFruits or berriesRoots or rhizomes	710.1
Malathion	SeedsFruits or berriesRoots or rhizomes	210.5
Metalaxyl	Seeds	5
Methamidophos	Entire Group 028	0.1
Parathion	SeedsFruits or berriesRoots or rhizomes	0.10.20.2
Parathion-methyl	SeedsFruits or berriesRoots or rhizomes	550.3
Permethrin	Entire Group 028	0.05
Phenthoate	Seeds	7
Phorate	SeedsFruits or berriesRoots or rhizomes	0.50.10.1
Phosalone	SeedsFruits or berriesRoots or rhizomes	223
Pirimicarb	Seeds	5
Pirimiphos-methyl	Seeds sub groupFruits sub group	30.5
Quintozene	Seeds sub groupFruits or berriesRoots or rhizomes	0.10.022
Vinclozolin	Entire spice group	0.05

European Union

In September 2008, the European Union issued new and revised Maximum Residue Limits (MRLs) for the roughly 1,100 pesticides ever used in the world. The revision was intended to simplify the previous system, under which certain pesticide residues were regulated by the

Commission; others were regulated by Member States, and others were not regulated at all.

New Zealand

Food Standards Australia New Zealand develops the standards for levels of pesticide residues in foods through a consultation process. The New Zealand Food Safety Authority publishes the maximum limits of pesticide residues for foods produced in New Zealand.

United Kingdom

Monitoring of pesticide residues in the UK began in the 1950s. From 1977 to 2000 the work was carried out by the Working Party on Pesticide Residues (WPPR), until in 2000 the work was taken over by the Pesticide Residue Committee (PRC). The PRC advise the government through the Pesticides Safety Directorate and the Food Standards Agency (FSA).

United States

In the US, tolerances for the amount of pesticide residue that may remain on food are set by the EPA, and measures are taken to keep pesticide residues below the tolerances. The US EPA has a web page for the allowable tolerances. In the US, the Food and Drug Administration (FDA) and USDA also routinely check food for the actual levels of pesticide residues.

Japan

In Japan, pesticide residues are regulated by the Food Safety Act. Pesticide tolerances are set by the Ministry of Health, Labour and Welfare through the Drug and Food Safety Committee. Unlisted residue amounts are restricted to 0.01ppm.

China

In China, the Ministry of Health (China) and the Ministry of Agriculture (China) have jointly established mechanisms and working procedures relating to maximum residue limit standards, while updating them continuously, according to the Food safety Law and regulations issued by the State Council. From GB25193-2010 to GB28260-2011, from Maximum Residue Limits for 12 Pesticides to 85 pesticides, they have improved the standards in response to Chinese national needs.

Health Impacts

Pesticide exposures can produce two distinct types of adverse health effects – acute (short-term) effects and chronic (long-term) effects. The severity of adverse health effects are determined by following factors:

the dose, the route of exposure, an individual's genetic vulnerability, age at the time of exposure and general health condition; the length of exposure, environmental factors and intakes with other chemicals through other factors.

Chronic Impacts

Possible symptoms of Long term pesticide exposure are listed as follows.

- Increased risk of cancer
- Neurological impairment (e.g. Parkinson's disease)
- Developmental effects
- Reproductive effects
- Organ damage
- Intrusion with the human hormone system

Many pesticides achieve their intended use of killing pests by disrupting the nervous system. Due to similarities in brain biochemistry among many different organisms, there is much speculation that these chemicals can have a negative impact on humans as well. There are also studies that show positive correlations between exposure to pesticides through occupational hazard, which tends to be significantly higher than that ingested by the general population through food, and the occurrence of certain cancers. Although the most of the general population may not exposed to large portion of pesticides, many of the pesticide residues that are attached tend to be lipophilic and can bio-accumulate in the body. Concerns have been raised about the possible role of continuous low-dosage exposure in causing certain cancers.

On the other hand, pesticide use has proven to be a valuable tool in combating disease vectors that spread illness to millions of children and adults each year. They are used in many developing countries to prevent the spread of malaria, leishmaniasis, dengue, and Japanese encephalitis which are diseases that have an enormous economical burden on society.

Acute Impacts

Some health effects from pesticide exposure may occur immediately following exposure. Symptoms of immediate body reaction against pesticide are as follows:

- Eyes: tearing, irritation, conjunctivitis
- Skin: rash, blistering, burns, sweating, contact dermatitis, jaundice

- Nervous system: headache, dizziness, mood disturbances, depression, stupor, muscle twitching, lack of coordination, seizures, paralysis, loss of consciousness, coma
- Respiratory system: throat pain, rhinorrhea, cough, pulmonary edema, difficulty in breathing, respiratory failure
- Cardiovascular system: cardiac arrhythmias
- Gastrointestinal tract: nausea, vomiting, diarrhea, abdominal pain

Acute poisoning can happen in various situations; by consuming food containing pesticide residue, including the accidental ingestion in the home, garden, farm or forest. Children are the most vulnerable to poisoning as they have not yet built up antibodies to certain toxicities.

Pesticide Poisoning

In August 1994, a serious incident of pesticide poisoning of sweet potato crops occurred in Shandong province, China. Because local farmers were not fully educated in the use of insecticides, they used the highly-toxic pesticide named parathion instead of trichlorphon. It resulted in over 300 cases of poisoning and 3 deaths. Also, there was a case where a large number of students were poisoned and 23 of them were hospitalized because the vegetables they consumed contained pesticide residues.

Child Neurodevelopment

Children are thought to be especially vulnerable to exposure to pesticide residues, especially if exposure occurs at critical windows of development. Infants and children consume higher amounts of food and water relative to their body-weight have higher surface area (i.e. skin surface) relative to their volume, and have a more permeable blood-brain barrier, and engage in behaviours like crawling and putting objects in their mouths, all of which can contribute to increased risks from exposure to pesticide residues through food or environmental routes. Neurotoxins and other chemicals that originate from pesticides pose the biggest threat to the developing human brain and nervous system. Presence of pesticide metabolites in urine samples have been implicated in disorders such as attention deficit hyperactivity disorder (ADHD), autism, behavioural and emotional problems, and delays in development. There is a lack of evidence of a direct cause-and-effect relationship between long-term, low-dose exposure to pesticide residues and neurological disease, partly because manufacturers are not always legally required to examine potential long-term threats.

Residues on Food

"Dirty Dozen" and "Clean Fifteen"

According to the Environmental Working Group choosing to buy certain organic food can significantly lower residue exposure by as much as about 90%. The group came up with a list of fruits and vegetable that are recommended to avoid as they are known to contain high levels of pesticide residues.

Even though the dirty dozen foods may have more pesticide residue than other produce, their level of residue is still very small compared to doses actually found to have an effect with chronic low-level exposure. A study calculated predicted long term consumer exposure to these pesticides; it was found exposure levels were 1000 times less than the lowest levels shown to have an effect. The scientific evidence shows that there is a very low risk associated with eating these foods despite their dubious label.

These foods, referred to as the "Dirty Dozen" readily absorb pesticides and retain them, so even if washed, they will still have noticeable levels of residues. The ability of a fruit or vegetable to retain or repel pesticides is dependent on its outer shell or protective layer. Fruits such as strawberries are more likely to absorb and retain pesticides than pineapples are.

Dirty Dozen

- apples
- celery
- cherry tomatoes
- cucumbers
- grapes
- hot peppers
- nectarines (imported)
- peaches
- potatoes
- spinach
- strawberries
- sweet bell peppers
- plus collard greens and kale
- plus summer squash and zucchini

Clean Fifteen

- asparagus
- avocados
- cabbage
- cantaloupe
- corn
- eggplant
- grapefruit
- kiwi fruit
- mango
- mushrooms
- onions
- papayas
- pineapples
- sweet peas (frozen)
- sweet potatoes

Minimizing Exposure to Pesticide Residues

Organic vs Non-Organic

Even though some organic foods contain significantly less amounts of pesticides than non-organically produced products, they still contain certain amounts of residue levels that are persistent in the environment. The methods of organic farming prohibit the use of many pesticides, however certain chemicals could be persistent in the soil, reach the organic produce via pesticide drift from nearby farms, or contaminate produce during transportation to the consumer. In addition to choosing to buy organic produce, there are certain measures that can be taken to minimize exposure to harmful residues: vegetables washes may help remove chemical residues, peeling fruits and vegetables that have known high levels of pesticide residues, and growing your own essential produce.

11

Miridae

The large and diverse insect family Miridae contains the plant bugs, leaf bugs, and grass bugs, and may also be known as capsid bugs. It is the largest family of true bugs belonging to the suborder Heteroptera, with over 10,000 known species and new ones constantly being described. They are small, terrestrial insects, usually oval-shaped or elongate and measuring less than 12 millimetres (0.5 in) in length. Some are brightly coloured, others drab or dark. Some genera are ant mimics at certain stages of life. Most of the more well-known mirids have received attention because they are agricultural pests. They pierce plant tissues and feed on the juices.

Lygus Bugs (*Lygus* spp.)

The genus *Lygus* includes over 40 species of plant-feeding insects in the family Miridae. At one time, nearly 200 species were classified as genus *Lygus*, but most of those have since been reclassified into new or existing genera. The term lygus bug is used for any member of genus *Lygus*. The more well-known lygus bugs are those that have agricultural impacts. Some lygus bugs are very serious agricultural pests.

The insects appear as small oval creatures. Adult lygus are approximately 3 mm wide and 6 mm long, coloured anything in a range from pale green to reddish brown or black. The bugs can be solid shaded or mottled, and have a distinctive triangle or V-shape on their backs. Adults are capable of flight, and will often thus escape when approached. Nymphs are wingless, and being light green in colour, are often mistaken for aphids. However, lygus nymphs have harder shells, are typically more active, gain spots as they age, and lack aphid cornicles.

Lygus bugs are known for their destructive feeding habits - they puncture plant tissues with their piercing mouthparts, and feed by sucking sap. Both the physical injury and the plant's own reaction to the bugs' saliva cause damage to the plant. The females insert their eggs directly into the plant tissues using piercing ovipositors, and the newly emerged nymphs are voracious consumers of plant tissue juices. Signs that a plant has been attacked by lygus bugs include discolouration, deformation of shoots and stems, curling of leaves, and lesions on the plant tissues.

- The tarnished plant bug (*Lygus lineolaris*) feeds on over half of all commercially-grown crop plants, but favours cotton, alfalfa, beans, stone fruits, and conifer seedlings. This bug can be found across North America, from northern Canada to southern Mexico.
- The western tarnished plant bug (*Lygus hesperus*) is a very serious pest of cotton, strawberries, and seed crops such as alfalfa. In the state of California alone the bug causes US$30 million in damage to cotton plants each year, and at least US$40 million in losses to the state's strawberry industry.
- The European tarnished plant bug (*Lygus rugulipennis*) is distributed throughout Europe, where it will feed on over 400 types of crop plant from peach trees to wheat to lettuce.

Some methods of biological pest control have proved useful against lygus bugs. For example, wasps of the genus *Peristenus* are parasitoids of lygus bugs; an adult wasp will inject an egg into a lygus nymph, and once the egg hatches the wasp's larva will consume the nymph from the inside out.

Stenotus Binotatus, a Minor Pest of Cereal Crops, Especially Wheat

Stenotus binotatus is a species of plant bug, originally from Europe, but now also established across North America and New Zealand. It is 6–7 mm (0.24–0.28 in) long, yellowish, with darker markings on the pronotum and forewings. It feeds on various grasses, and can be a pest of crops such as wheat.

Description

Stenotus binotatus *is a fairly large plant bug (6–7 millimetres or 0.24–0.28 inches long), which is somewhat variable in appearance. The insect's sides are roughly parallel, and the colours depend on both the animal's sex and its age, the markings becoming darker and stronger with increasing age. Males are mostly yellow, with darker markings on the pronotum and forewings, which females are greenish-yellow with paler markings.*

Distribution

S. binotatus *is native to Europe but has been introduced to temperate regions around the world. It is common "throughout the northern and central U.S. and southern Canada", and it has been introduced to New Zealand, where it now found almost throughout the country.*

Ecology

*Both the nymphs and adults feed on the inflorescences of grasses, especially timothy-grass (*Phleum pratense*). In New Zealand,* S. binotatus *has been collected from a large variety of plants, including the grasses brown top, Yorkshire fog, cocksfoot, wheat, barley and maize, and many other plants (including rushes,* Hypericum*, mânuka, râtâ,* Nothofagus, Coriaria, Olearia, Muehlenbeckia, Carmichaelia, Larix decidua, Eucalyptus, Melicytus ramiflorus, Coprosma robusta*, pôhutukawa, hemlock and nettles).*

Adults are active from June to September in the United Kingdom, and in December and January in New Zealand.

Crop Damage

Stenotus binotatus *produces an enzyme which degrades the gluten in the grasses it feeds on. These enzymes are believed to be responsible for the characteristic "sticky dough" produced from wheat which has been attacked by plant bugs such as* S. binotatus.

The wheat grain does not show a pale area or puncture mark after attack by S. binotatus*, in contrast to other bugs. Instead, the grains appear collapsed. Wheat grains from known bug infestation in New Zealand do not resemble those produced by* S. binotatus*, and SDS-PAGE patterns also suggest that* S. binotatus *was not the cause of infestations.*

Taxonomic History

Stenotus binotatus *was originally described under the name* Lygaeus binotatus *by Johan Christian Fabricius in 1794. The genus* Stenotus *was erected by Wassily Ewgrafowitsch Jakowlew (also transcribed as "Jakovlev") in 1877, and* S. binotatus *was designated its type species.*

S. binotatus *is known by various common names, including* two-spotted plant bug, timothy plant bug, *and* slender crop mirid.

Apple Dimpling Bug (Yellow Mirid)

The apple dimpling bug (yellow mirid), *Campylomma liebknechti* (Girault) is about one third the size of the green mirid. It is an important pest of apples and is known to damage small squares early in squaring on cotton but it is generally regarded more as a predator of heliothis eggs and mites than a pest.

The threshold for the apple dimpling bug is generally about 5 times greater than the threshold for the green mirid. If the apple dimpling bug is in high numbers it is worthwhile to monitor the plants fruit set by either checking fruit retention or by using the 'fruiting factor' technique. This will indicate if damage is occuring which may require control.

Mosquito Bugs

The genus *Helopeltis*, also known as mosquito bugs, is a group of Heteropterans in the family Miridae (capsid bugs). They include important pests of various crops, including cacao, cashew, cotton and tea. Now in a different subgenus, a number of similar *Afropeltis* species are pests in Africa. Mosquito bugs have a characteristic spine on the scutellum, which is a diagnostic feature. Classification in the field is based on morphological characteristics, with considerable variations in colouration between insects of the same species (although for example, *H. theivora* is characteristically green and *H. antonii* red-brown).

Damage and Distribution

With typical Hemipteran sucking mouthparts, they pierce plant tissues and cause damage ranging from leaf tattering and fruit blemishes, to complete death of shoots, branches or whole plants. There has been evident speciation along the islands of the Malay archipelago and there may be cryptic species in this genus. Helopeltis spp. sensu stricto are important 'new encounter' pests of SE Asian cocoa including:

- *H. antonii*: India through to West Irian
- *H. bakeri*: Malay peninsula and Philippines
- *H. bradyi*: Sri Lanka, Malaysia, Indonesia
- *H. clavifer*: Sabah and Papua New Guinea
- *H. collaris*: Philippines
- *H. sulawesi*: Sulawesi
- *H. sumatranus*: Sumatra
- *H. theivora* (with *H. theobromae* as a sub-species): India through to Java

Afropeltis and Other Cocoa Mirid Pest Species

Very similar to Helopeltis, *the African species were placed into* Afropeltis *by Schmitz (1968); these species usually have a lesser pests status than their Asian counterparts, with* Sahlbergella singularis *and* Distantiella theobromae *causing greatest cocoa tree and crop damage.* Monalonion *species, belonging to the same tribe, are similarly minor pests of Latin American cocoa.*

Honeylocust Plant Bug (*Diaphnocoris Chlorionis*)

Blepharidopterus chlorionis, known commonly as the honeylocust plant bug and under the synonym *Diaphnocoris chlorionis*, is a hemipteran in the family Miridae. They are a major pest of the honey locust tree (*Gleditsia tricanthos*), a deciduous tree native to central North America.

The honeylocust plant bug is a key pest of honeylocust, *Gleditsia tricanthos* . It is a fairly recent problem, not becoming a concern until the 1950s and 1960s with the introduction of the thornless cultivars often grown in nurseries and used in landscapes. The insect can be found throughout the eastern United States; anywhere honeylocust is growing naturally or in the landscape.

Description

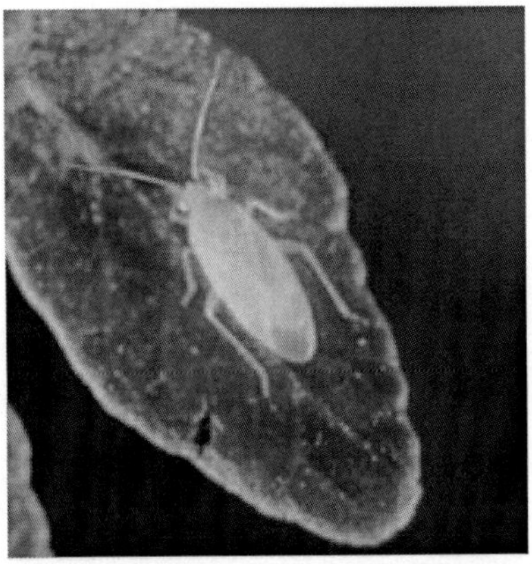

***Figure:**. Honeylocust plant bug (adult)*

This key pest is a plant bug that belongs to the insect family Miridae. This species is easily overlooked because the colour of the nymph and adult closely resembles the new growth of the tree. The nymph looks nearly identical to the adult in colour and form, except for its smaller size and shortened wing buds. The nymph has a distinct orange spot in the middle of the abdomen. Both stages of the insect are rather active and adults will fly when disturbed. Adults are approximately 3/16 inch long, and light green to yellow. The insect has beaklike mouthparts that are used to remove fluids from plant cells. The overwintering egg is light coloured, resembling a grain of rice.

Life History

Adult females lay linear clusters of eggs just under the bark of 2- to 3-year-old twigs in midsummer. Eggs begin to hatch just after the vegetative buds of honeylocust begin to break in early spring. After eggs hatch in late April, nymphs undergo five instars until maturing into adults in early June. Nymphal development requires approximately 30 days.

Populations are most abundant during early June, slowly declining in late June, and disappearing by early July. This pest has one generation per season.

Damage

***Figure:**. Honeylocust plant bug (damage)*

This plant bug species is a fairly unpredictable pest when trying to prepare for an infestation. Both nymphs and adults can feed on the foliage with piercing-sucking mouthparts, although nymphs cause the most damage. In early spring young nymphs crawl into unfolding leaves and begin feeding just after bud break. Damage is caused when the insect injects toxic saliva during feeding.

The toxin kills cells around the feeding site causing a yellow spot that eventually turns brown. Feeding could result in leaf rolling, distortion, stunting, and chlorosis. Heavy infestations may cause complete defoliation of the host, although tree death rarely occurs. The most common damage is leaf distortion, discolouration, and dwarfed leaflets. This damage remains apparent throughout the growing season. The most susceptible cultivars seem to be those with yellow leaves such as 'Sunburst'. The green-leaved varieties such as 'Shademaster' and

'Skyline' are less susceptible to injury caused by this pest. Damage may weaken the tree and leave it susceptible to attack by secondary insects and diseases.

Management

Newly emerging leaves should be monitored in the spring for symptoms of stippling and distortion. The use of a black beating tray in the spring will assist in determining the presence of the young, nymphal stages of this pest.

Non-Chemical

Nymphs may be knocked off leaves with a high-pressure spray of water. Planting green-leaved cultivars that are less susceptible to attack by this pest than the yellow-leaved cultivars will reduce the risk of an infestation of this pest.

Chemical

Large established trees can withstand multiple years of damage from this pest without serious harm. However, insecticidal soap and horticultural oil may be applied according to label directions in early spring after nymphs appear and when new leaflets are forming. Conventional insecticides should be applied according to label directions in the spring to prevent leaf damage. To effectively manage this pest, registered insecticides should target young nymphs 7 to 10 days after bud break in late April to early May. It is important that the foliage and bark are completely covered. Chemical control may not always be effective because this pest is highly mobile and individuals may migrate from untreated areas

Warning

Pesticides are poisonous. Read and follow directions and safety precautions on labels. Handle carefully and store in original labelled containers out of the reach of children, pets, and livestock. Dispose of empty containers right away, in a safe manner and place. Do not contaminate forage, streams, or ponds.

Green Mirid (*Creontiades Dilutus*) Damages Many Types of Field Crops

Mirids can cause significant crop losses in mung-beans with yield reductions of up to 25–50% common where high mirid populations (eg 10/m2) are left uncontrolled. Mirids can reduce yields by 60 kg/ha for every mirid/m2 of crop. Mirids can be present in mung-beans at any stage from seedlings to podding. Budding, flowering and early-podding crops are at greatest risk. Low populations of green mirids are often present in vegetative crops but there is no evidence they cause 'tipping' of vegetative terminals or have any impact on yield. Mirid populations usually increase with the onset of budding and peak during late pod-filll. Generally around 80% of mirids present in flowering/podding crops are nymphs..

Damage Caused by Mirids

Mirids are sucking insects and feed by piercing the plant tissue and releasing a chemical that destroys cells in the feeding zone. This causes plant tissue to discolour and die. Mirids prefer to feed on flowers, buds and young pods, causing these to abort (shed).

Mirid damage reduced mungbean pods (above)

Severe mirid damage to pods (right)

Severe mirid damage in mungbean results in fewer harvestable pods being set, however similar symptoms can also be caused by thrips, high temperatures and moisture stress. Mirids may also attack more mature pods, damaging the seeds inside without causing shedding. If pod set is greatly reduced and mirid populations are relatively low (e.g. 0.5 — 1.0 /m2) — loss may be due to other factors.

Mung-beans always set many more buds and flowers than they can convert to harvestable pods. Typically (in dryland crops) only 33% of buds/flowers are converted to harvestable pods. This % may be even lower in moisture stressed crops.

Recent Helicoverpa threshold trials show that mungbeans can compensate for considerable early damage (at budding/flowering) of up to 80% bud loss with no yield loss. However this only occurs when there is adequate moisture and no subsequent pest damage and there will be a delay in crop maturity. . The loss of replacement buds through continuous pest pressure will lead to reduced crop yields.

Medium and large mirid nymphs (instars 3–5) are as damaging as adults. Although small mirid nymphs (instars 1–2) are less damaging, they soon develop into larger more damaging pests (within 3–5 days respectively at 300C).

Identification of Mirids

Three mirid species attack mungbeans and include green, brown and (less frequently) crop mirids. All these species are equally damaging.

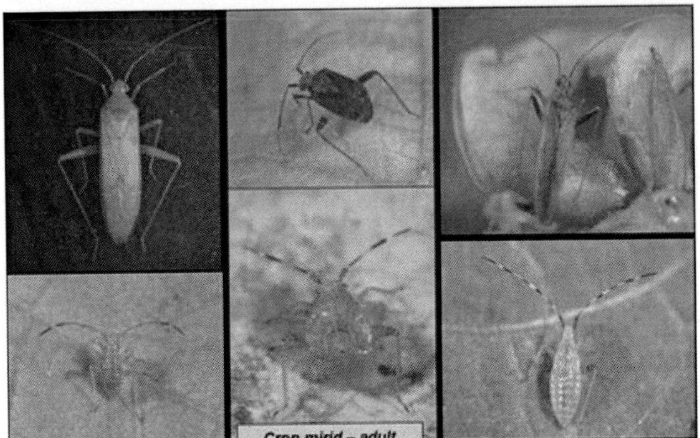

Green mirid – adult (top), nymph (bottom)

Crop mirid – adult (top), nymph (bottom)

Brown mirid – adult (top), nymph (bottom)

Where do the Mirids Come from?

Mirids move into summer crops when alternative host plants dry off and they seek a fresh food source. Lucerne and sunflowers can be important sources of mirids in mungbean growing areas.

Influxes of mirid adults often coincide with northwest winds in spring. There is also evidence of long distance migration, possibly from inland areas, associated with weather fronts. This may be the cause of some of the widespread and repeated influxes of mirids sometimes observed in grain and cotton growing regions early in the summer.

Monitoring for Mirids

- Mirids are very mobile and in-crop populations can increase rapidly from budding onwards.
- Crops should be inspected weekly during the vegetative stage (to pick the start of budding) and twice weekly from budding onwards until post flowering.
- In row crops, the preferred method is beat sheeting, as this method is the most effective for helicoverpa and pod-sucking bugs.

- Sample 5 one-metre lengths of row (not consecutive) within a 20 m radius, from at least 6 sites throughout a crop.
- Avoid sampling during very windy weather as mirids are easily blown off the sheet.

Action Level

Mirid thresholds for budding/flowering mungbeans are 0.3–0.6 per m2 for aerial and ground rig respectively. While mirid thresholds are very low, this is more a reflection of the cheapness of the preferred mirid pesticide – dimethoate ($4/ha + application). The above thresholds are based on the cost of dimethoate. Were indoxacarb to be used, the threshold for an aerial sprayed crop would be 1 mirid/m2.

Note that these thresholds are the break even point (cost of control = value of likely damage) and that action only needs to be taken if populations exceed these levels. However in practice, mirid populations usually increase rapidly from budding onwards and spraying at-threshold populations would be justified. Because of the crop's ability to compensate for early damage, spraying low mirid populations (i.e. at threshold) at very early budding can be delayed slightly with no risk to yield or harvest maturity. Taking this approach reduces the need for a subsequent mirid spray.

Cultural Control

A crop that has a short flowering period reduces the risk of mirid damage. Flowering periods can be shortened by planting on a full moisture profile and by watering crops just before budding. Consider planting crops in at least 50 cm rows (as opposed to broadcast planting) to facilitate easier pest sampling.

Chemical Control

- Dimethoate at 500 mL/ha (all summer pulses) or
- Indoxacarb at 400 mL/ha (mungbeans and soybeans only).

Dimethoate is often applied at lower than label rates (eg 200–250 mL/ha). These rates give excellent mirid control but have far less impact on many beneficials. Trail results have shown that the addition of salt (0.5% NaCl) as an adjuvant, improves the effectiveness of dimethoate at lower rates. The amount of salt used (0.5%) has no phytotoxic effect on summer pulse crops. 'Hard' water can markedly lower the effectiveness of dimethoate and should be countered by adding a buffering agent such as LI700.

Indoxacarb is not recommended at high mirid pressure (>2/m2) because it is less effective than dimethoate. Because it is restricted to one (1) spray application per crop, indoxacarb is best reserved for Helicoverpa control during podding. As mentioned previously, the mirid thresholds are considerably higher if indoxacarb is used.

Natural Enemies

There are no beneficial species that are recognised to be regulators of mirid populations in mungbeans. However damsel bugs, big-eyed bugs, predatory shield bugs, as well as lynx, night stalker and jumping spiders are known to feed on mirid adults, nymphs and eggs. Naturally occurring fungi (eg Beauvaria) may also infect and kill mirids, but are rarely observed in the field.

Potato Mirid (*Calocoris Norvegicus*)

Closterotomus norvegicus (also known as the potato capsid) is a species of bugs belonging to the family Miridae, subfamily Mirinae. It can be found feeding on nettle, clover, and cannabis, as well as Compositae, potatoes, carrots and chrysanthemums. They prefer to feed on the flowers, buds and unripe fruit. The species occurs in Great Britain and continental Europe. In 1997 it was moved from the *Calocoris* genus to its current name.

Description

The species is green coloured as a nymph but when they get to adulthood they colour changes to reddish-brown. It does however, vary by territory, for example specimens from northern Britain are brownish-black. The prothorax have two spots while its scutellum have dark marks.

How to Recognise it

Capsid bugs are small (up to 6mm long) insects which vary from pale yellow and green to red/brown in colour. The adults are winged but the nymphs are wingless. You may spot the bugs on the plant, but they are very timid and usually drop to the ground or fly away when disturbed. The damage caused usually appears as small, ragged, brown-edged holes appearing in young leaves, which then become tattered as the independent holes 'join up'. Leaves can also appear distorted or 'puckered'. Damage may be seen on buds and shoots.

Lifecycle

The lifecycle varies between species. Eggs hatch in spring and nymphs feed from late spring through summer, maturing into the winged adults. There may be more than one generation each year. Capsid bugs overwinter either as eggs or adults (depending on the species).

Why it's a Problem

Both the nymphs and adults feed off the sap of plants, piercing the plant and injecting saliva, which kills plant tissues (causing the irregular holes), before sucking out the sap. This results in the eventual disintegration of affected leaves, the death of buds and shoots, and any flowers developing from affected buds being deformed. Fruit produced (eg apples) can have bumps or other irregularities as a result of the infestation, but is still edible.

Where You are Likely to Find it

The damage from capsid bugs is generally most noticeable from late spring through summer and they can attack a very wide range of garden plants. Some of the principle plants affected include apple trees, currants, hawthorns, potatoes, pears, plums, gooseberries, rapsberries, strawberries, runner beans, asters, arctotis, buddlejas, caryopteris, ceanothus, chrysanthemums, dahlias, forsythias, fuchsias, hydrangeas, magnolias, nasturtiums, pelargoniums, poppies, roses, salvias, sunflowers, venidiums and zinnias.

How to Deter it

Clear up debris and fallen leaves in winter, particularly under hedges, which may be providing overwintering accommodation for the eggs or adults. Keeping weeds to a minimum will also remove any alternative hosts for this pest.

How to Get Rid of it

This isn't usually serious enough to merit chemical treatment (particularly not on vegetables), but if the damage is particularly unsightly or affecting soft fruit crops, an insecticide such as pyrethrin, a synthetic pyrethroid compound such as bifenthrin, or thiacloprid can be used on the plant and surrounding ground (to include bugs which drop off the plant when disturbed).

This can be done on woody plants (which may harbour overwintering eggs) in December or January when the plants are dormant, but this will not protect the plants from re-invasion in the spring, so spraying these, and herbaceous plants, in May, June and July when the pests are active will be more effective.

Spray strawberries, raspberries and loganberries immediately before flowering and apples, pears, plums, currants and gooseberries immediately after flowering. Ornamental plants should be sprayed in spring and summer and may need a further application later in the year.

Is it Good for Anything?!

Some species of capsid bugs feed on other creatures which are considered pests, including caterpillars, mites and aphids.

Common Green Capsid - *Lygocoris Pabulinus*

How to Recognise it

The adult is bright green and measures about 5mm in length, resembling a large aphid. The nymphs (similar to the adults but without wings) are yellow-green in colour. Careful inspection may reveal capsids, but they quickly drop to the ground or fly away when disturbed so you may never see them.

Foliage has distorted growth with small brown spots which become holes, turning raggedy edged as the leaf expands. Buds and shoots can be damaged, turning brown and withering or causing flowers to open lop-sided. Fruit can be scarred with bumps and other irregularities.

Most of the damage is done in late spring and summer.

Lifecycle

The adults lay eggs on young twigs of woody host plants (eg apples, currants, hawthorns) in autumn, which overwinter and hatch as nymphs in the spring. The nymphs feed on the new growth and then migrate to herbaceous plants for the summer. The females mature on the herbaceous plants and lay eggs in the stems and petioles during June and July. This second generation hatches and matures to adulthood in autumn, the females laying eggs to overwinter before dying.

Why it's a Problem

Its salivary juices (which it injects into the plant when feeding on the sap) are poisonous, causing the telltale leaf, bud and fruit distortions.

Where you are Likely to Find it

On fruit trees, shrubs and flower crops (such as dahlias and chrysanthemum), usually outside. The adult capsid can fly from plant to plant.

How to Deter it

Clear up debris and fallen leaves in winter, particularly under hedges, which may be providing overwintering accommodation for the eggs.

How to Get Rid of it

This isn't usually serious enough to merit chemical treatment, but if the damage is particularly unsightly or affecting soft fruit crops, an insecticide such as pyrethrin, a synthetic pyrethroid compound such as bifenthrin, or thiacloprid can be used on the plant and surrounding ground. This can be done in December or January when the plants are dormant, but this will not protect the plants from re-invasion in the spring, so spraying in May, June and July when the pests are active will be more effective. Spray strawberries, raspberries and loganberries immediately before flowering and apples, pears, plums, currants and gooseberries immediately after flowering. Ornamental plums should be sprayed in spring and summer and may need a further application later in the year.

Is it Good for Anything?!

Yes, the common green capsid feeds on mites, aphids, caterpillars and other small invertebrates which are considered garden pests.

Deraeocoris Nebulosus

Deraeocoris nebulosus (Uhler) is a generalist predator of plant-feeding insects and mites. It is associated with many common pests on more than 50 species of ornamental trees and shrubs. *D. nebulosus* is found in southern Canada and is widespread in the United States; it is common in the eastern states. Reported by Uhler to be predaceous on a cankerworm, it may have been the first mirid documented as a predator in North America.

Appearance

Adults have ovate, shiny, dark, olive bodies with pale markings, and are 3.5-4.0 mm long and 1.75-2.0 mm wide. The apical half of each forewing is clear with each having a small fuscous dot which helps distinguish *D. nebulosus* from other species in the genus. Eggs (0.91 mm long, 0.28 mm wide) are laid singly or in groups of two or more in leaf midveins and petioles, with only the operculum, including a respiratory horn, visible. Nymphs are pale grey, with the early instars having a red tinge and the late instars having red streaks on the legs and a red line between two of the segments on the abdominal dorsum.

Habitat

Deraeocoris nebulosus, collected on more than 75 plant species, is found in apple orchards, peach orchards, pecan orchards, cotton fields and in landscape plantings.

Pests Attacked (Host Range)

This predator feeds on whiteflies, aphids, psyllids, scales, mites, and lace bugs. In captivity it tends to cannibalize unless provided with hiding places.

Life Cycle

Adults overwinter in protected places such as under bark. With onset of warmer weather and appearance of new leaves, eggs are laid in the petioles and midveins. This bug has five nymphal stages. Under laboratory conditions, the nymphal period lasts 19.8 days at 21°C and 13.3 days at 27°C. Females have a mean fecundity of about 240 eggs. Three generations per year are reported from Pennsylvania, with more suspected farther south. Adults track a succession of plant species, according to prey availability.

Relative Effectiveness

Wheeler et al. reported that *D. nebulosus* consumed an average of 107.6 lace bug nymphs during development and 6.9 nymphs per day as an adult. An adult can consume 4-7 cotton aphids per day and 16-19 tobacco budworm eggs per day. Little work has been done with *D. nebulosus* in the field.

Conservation

Use of target-specific pesticides only after a pest reaches an economic threshold will help conserve *D. nebulosus*.

Pesticide Susceptibility

Little work has been done on the susceptibility of *D. nebulosus* to pesticides. Synthetic pyrethroids show a low toxicity, only up to 10% and 30% susceptible at the low and high rates tested, respectively.

Deraeocoris brevis, a species common in the northwestern United States, is susceptible to many pesticides including azinphosmethly, fenvalerate, diflubenzuron, and organophosphates

Insects and Insect Relatives

Insects (from Latin *insectum*, a calque of Greek —íôïìïí [*éntomon*], "cut into sections") are a class of invertebrates within the arthropod phylum that have a chitinous exoskeleton, a three-part body (head, thorax and abdomen), three pairs of jointed legs, compound eyes and one pair of antennae. They are among the most diverse groups of animals on the planet, including more than a million described species and representing more than half of all known living organisms. The number of extant species is estimated at between six and ten million, and potentially represent over 90% of the differing animal life forms on Earth. Insects may be found in nearly all environments, although only a small number of species reside in the oceans, a habitat dominated by another arthropod group, crustaceans.

The life cycles of insects vary but most insects hatch from eggs. Insect growth is constrained by the inelastic exoskeleton and development involves a series of molts. The immature stages can differ from the adults in structure, habit and habitat, and can include a passive pupal stage in those groups that undergo complete metamorphosis. Insects that undergo incomplete metamorphosis lack a pupal stage and adults develop through a series of nymphal stages. The higher level relationship of the hexapoda is unclear. Fossilized insects of enormous size have been found from the Paleozoic Era, including giant dragonflies with wingspans of 55 to 70 cm (22–28 in). The most diverse insect groups appear to have coevolved with flowering plants.

Adult insects typically move about by walking, flying, or sometimes swimming. As it allows for rapid yet stable movement, many insects adopt a tripedal gait in which they walk with their legs touching the ground in alternating triangles. Insects are the only invertebrates to have evolved flight. Many insects spend at least part of their lives under water, with larval adaptations that include gills, and some adult insects are aquatic and have adaptations for swimming. Some species, such as water striders, are capable of walking on the surface of water. Insects are mostly solitary, but some, such as certain bees, ants and termites, are social and live in large, well-organised colonies. Some insects, such as earwigs, show maternal care, guarding their eggs and young. Insects can communicate with each other in a variety of ways. Male moths can sense the pheromones of female moths over great distances. Other species communicate with sounds: crickets stridulate, or rub their wings together, to attract a mate and repel other males. Lampyridae in the beetle order Coleoptera communicate with light.

Humans regard certain insects as pests, and attempt to control them using insecticides and a host of other techniques. Some insects damage crops by feeding on sap, leaves or fruits. A few parasitic species are pathogenic. Some insects perform complex ecological roles; blow-flies, for example, help consume carrion but also spread diseases. Insect pollinators are essential to the life-cycle of many flowering plant species on which most organisms, including humans, are at least partly dependent; without them, the terrestrial portion of the biosphere (including humans) would be devastated. Many other insects are considered ecologically beneficial as predators and a few provide direct economic benefit. Silkworms and bees have been used extensively by humans for the production of silk and honey, respectively. In some cultures, the larvae or adults of certain insects are a food-source for humans.

Phylogeny and Evolution

Figure: *Evolution has produced astonishing variety in insects. Pictured are some of the possible shapes of antennae.*

The evolutionary relationship of insects to other animal groups remains unclear. Although more traditionally grouped with millipedes and centipedes, evidence has emerged favouring closer evolutionary ties with crustaceans. In the Pancrustacea theory, insects, together with Remipedia and Malacostraca, make up a natural clade. Other terrestrial arthropods, such as centipedes, millipedes, scorpions and spiders, are sometimes confused with insects since their body plans can appear similar, sharing (as do all arthropods) a jointed exoskeleton. However, upon closer examination their features differ significantly; most noticeably they do not have the six-leg characteristic of adult insects.

The higher-level phylogeny of the arthropods continues to be a matter of debate and research. In 2008, researchers at Tufts University uncovered what they believe is the world's oldest known full-body impression of a primitive flying insect, a 300 million-year-old specimen from the Carboniferous Period. The oldest definitive insect fossil is the Devonian *Rhyniognatha hirsti*, from the 396-million-year-old Rhynie chert. It may have superficially resembled a modern-day silverfish insect. This species already possessed dicondylic mandibles (two articulations in the mandible), a feature associated with winged insects, suggesting that wings may already have evolved at this time. Thus, the first insects probably appeared earlier, in the Silurian period.

There have been four super radiations of insects: beetles (evolved ~300 million years ago), flies (evolved ~250 million years ago), moths and wasps (evolved ~150 million years ago). These four groups account for the majority of described species. The flies and moths along with the fleas evolved from the Mecoptera.

The origins of insect flight remain obscure, since the earliest winged insects currently known appear to have been capable fliers. Some extinct insects had an additional pair of winglets attaching to the first segment of the thorax, for a total of three pairs. As of 2009, there is no evidence that suggests that the insects were a particularly successful group of animals before they evolved to have wings.

Late Carboniferous and Early Permian insect orders include both extant groups and a number of Paleozoic species, now extinct. During this era, some giant dragonfly-like forms reached wingspans of 55 to 70 cm (22 to 28 in) making them far larger than any living insect. This gigantism may have been due to higher atmospheric oxygen levels that allowed increased respiratory efficiency relative to today. The lack of flying vertebrates could have been another factor. Most extinct orders of insects developed during the Permian period that began around 270 million years ago. Many of the early groups became extinct during the Permian-Triassic extinction event, the largest mass extinction in the history of the Earth, around 252 million years ago.

The remarkably successful Hymenopterans appeared as long as 146 million years ago in the Cretaceous period, but achieved their wide diversity more recently in the Cenozoic era, which began 66 million years ago. A number of highly successful insect groups evolved in conjunction with flowering plants, a powerful illustration of coevolution.

Many modern insect genera developed during the Cenozoic. Insects from this period on are often found preserved in amber, often in perfect condition. The body plan, or *morphology*, of such specimens is thus easily compared with modern species. The study of fossilized insects is called paleoentomology.

Evolutionary Relationships

Insects are prey for a variety of organisms, including terrestrial vertebrates. The earliest vertebrates on land existed 400 million years ago and were large amphibious piscivores. Through gradual evolutionary change, insectivory was the next diet type to evolve.

Insects were among the earliest terrestrial herbivores and acted as major selection agents on plants. Plants evolved chemical defences against this herbivory and the insects in turn evolved mechanisms to deal with plant toxins. Many insects make use of these toxins to protect themselves from their predators. Such insects often advertise their toxicity using warning colours. This successful evolutionary pattern has also been utilised by mimics. Over time, this has led to complex groups of coevolved species. Conversely, some interactions between plants and insects, like pollination, are beneficial to both organisms. Coevolution has led to the development of very specific mutualisms in such systems.

Taxonomy

Traditional morphology-based or appearance-based systematics has usually given Hexapoda the rank of superclass, and identified four groups within it: insects (Ectognatha), springtails (Collembola), Protura and Diplura, the latter three being grouped together as Entognatha on the basis of internalized mouth parts. Supraordinal relationships have undergone numerous changes with the advent of methods based on evolutionary history and genetic data. A recent theory is that Hexapoda is polyphyletic (where the last common ancestor was not a member of the group), with the entognath classes having separate evolutionary histories from Insecta.

Many of the traditional appearance-based taxa have been shown to be paraphyletic, so rather than using ranks like subclass, superorder and infraorder, it has proved better to use monophyletic groupings (in which the last common ancestor is a member of the group). The following represents the best supported monophyletic groupings for the Insecta.

Insects can be divided into two groups historically treated as subclasses: wingless insects, known as Apterygota, and winged insects, known as Pterygota. The Apterygota consist of the primitively wingless order of the silverfish (Thysanura). Archaeognatha make up the Monocondylia based on the shape of their mandibles, while Thysanura and Pterygota are grouped together as Dicondylia. It is possible that the Thysanura themselves are not monophyletic, with the family Lepidotrichidae being a sister group to the Dicondylia (Pterygota and the remaining Thysanura).

Paleoptera and Neoptera are the winged orders of insects differentiated by the presence of hardened body parts called sclerites; also, in Neoptera, muscles that allow their wings to fold flatly over the abdomen. Neoptera can further be divided into incomplete metamorphosis-based (Polyneoptera and Paraneoptera) and complete metamorphosis-based groups. It has proved difficult to clarify the relationships between the orders in Polyneoptera because of constant new findings calling for revision of the taxa. For example, Paraneoptera has turned out to be more closely related to Endopterygota than to the rest of the Exopterygota. The recent molecular finding that the traditional louse orders Mallophaga and Anoplura are derived from within Psocoptera has led to the new taxon Psocodea. Phasmatodea and Embiidina have been suggested to form Eukinolabia. Mantodea, Blattodea and Isoptera are thought to form a monophyletic group termed Dictyoptera.

It is likely that Exopterygota is paraphyletic in regard to Endopterygota. Matters that have had a lot of controversy include Strepsiptera and Diptera grouped together as Halteria based on a reduction of one of the wing pairs – a position not well-supported in the entomological community. The Neuropterida are often lumped or split on the whims of the taxonomist. Fleas are now thought to be closely related to boreid mecopterans. Many questions remain to be answered when it comes to basal relationships amongst endopterygote orders, particularly Hymenoptera.

The study of the classification or taxonomy of any insect is called systematic entomology. If one works with a more specific order or even a family, the term may also be made specific to that order or family, for example systematic dipterology.

Distribution and Diversity

Though the true dimensions of species diversity remain uncertain, estimates range from 1.4 to 1.8 million species. This probably represents less than 20% of all species on Earth, and with only about 20,000 new species of all organisms being described each year, most species likely will remain undescribed for many years unless species descriptions increase in rate. About 850,000–1,000,000 of all described species are insects. Of the 24 orders of insects, four dominate in terms of numbers of described species, with an estimated 600,000–795,000 species included in Coleoptera, Diptera, Hymenoptera and Lepidoptera. Almost as many species of beetles have been named as all other insects added together, or all other noninsects (plants and animals).

Comparison of the estimated number of species in the four most speciose insect orders			
	Described species	*Average description rate (species per year)*	*Publication effort*
Coleoptera	300,000–400,000	2308	0.01
Lepidoptera	110,000–120,000	642	0.03
Diptera	90,000–150,000	1048	0.04
Hymenoptera	100,000–125,000	1196	0.02

Morphology and Physiology

External

Insects have segmented bodies supported by exoskeletons, the hard outer covering made mostly of chitin. The segments of the body are organised into three distinctive but interconnected units, or tagmata: a head, a thorax and an abdomen. The head supports a pair of sensory antennae, a pair of compound eyes, and, if present, one to three simple eyes (or ocelli) and three sets of variously modified appendages that form the mouthparts. The thorax has six segmented legs—one pair each for the prothorax, mesothorax and the metathorax segments making up the thorax—and, none, two or four wings. The abdomen consists of eleven segments, though in a few species of insects, these segments may be fused together or reduced in size. The abdomen also contains most of the digestive, respiratory, excretory and reproductive internal structures. Considerable variation and many adaptations in the body parts of insects occur, especially wings, legs, antenna and mouthparts.

Segmentation

The head is enclosed in a hard, heavily sclerotized, unsegmented, exoskeletal head capsule, or epicranium, which contains most of the sensing organs, including the antennae, ocellus or eyes, and the mouthparts. Of all the insect orders, Orthoptera displays the most features found in other insects, including the sutures and sclerites. Here, the vertex, or the apex (dorsal region), is situated between the compound eyes for insects with a hypognathous and opisthognathous head. In prognathous insects, the vertex is not found between the compound eyes, but rather, where the ocelli are normally. This is because the primary axis of the head is rotated 90° to become parallel to the primary axis of the body. In some species, this region is modified and assumes a different name.

The thorax is a tagma composed of three sections, the prothorax, mesothorax and the metathorax. The anterior segment, closest to the head, is the prothorax, with the major features being the first pair of legs and the pronotum. The middle segment is the mesothorax, with the major features being the second pair of legs and the anterior wings. The third and most posterior segment, abutting the abdomen, is the metathorax, which features the third pair of legs and the posterior wings. Each segment is dilineated by an intersegmental suture. Each segment has four basic regions. The dorsal surface is called the tergum (or *notum*) to distinguish it from the abdominal terga. The two lateral regions are called the pleura (singular: pleuron) and the ventral aspect is called the sternum. In turn, the notum of the prothorax is called the pronotum, the notum for the mesothorax is called the mesonotum and the notum for the metathorax is called the metanotum. Continuing with this logic, the mesopleura and metapleura, as well as the mesosternum and metasternum, are used.

The abdomen: is the largest tagma of the insect, which typically consists of 11–12 segments and is less strongly sclerotized than the head or thorax. Each segment of the abdomen is represented by a sclerotized tergum and sternum. Terga are separated from each other and from the adjacent sterna or pleura by membranes. Spiracles are located in the pleural area. Variation of this ground plan includes the fusion of terga or terga and sterna to form continuous dorsal or ventral shields or a conical tube. Some insects bear a sclerite in the pleural area called a laterotergite. Ventral sclerites are sometimes called laterosternites. During the embryonic stage of many insects and the postembryonic stage of primitive insects, 11 abdominal segments are present. In modern insects there is a tendency toward reduction in the number of the abdominal segments, but the primitive number of 11 is maintained during embryogenesis. Variation in abdominal segment number is considerable. If the Apterygota are considered to be indicative of the ground plan for pterygotes, confusion reigns: adult Protura have 12 segments, Collembola have 6. The orthopteran family Acrididae has 11 segments, and a fossil specimen of Zoraptera has a 10-segmented abdomen.

Exoskeleton

The insect outer skeleton, the cuticle, is made up of two layers: the epicuticle, which is a thin and waxy water resistant outer layer and contains no chitin, and a lower layer called the procuticle. The procuticle is chitinous and much thicker than the epicuticle and has two layers: an outer layer known as the exocuticle and an inner layer known as the endocuticle. The tough and flexible endocuticle is built from numerous layers of fibrous chitin and proteins, criss-crossing each other in a sandwich pattern, while the exocuticle is rigid and hardened. The exocuticle is greatly reduced in many soft-bodied insects (e.g., caterpillars), especially during their larval stages.

Insects are the only invertebrates to have developed active flight capability, and this has played an important role in their success. Their muscles are able to contract multiple times for each single nerve impulse, allowing the wings to beat faster than would ordinarily be possible. Having their muscles attached to their exoskeletons is more efficient and allows more muscle connections; crustaceans also use the same method, though all spiders use hydraulic pressure to extend their legs, a system inherited from their pre-arthropod ancestors. Unlike insects, though, most aquatic crustaceans are biomineralized with calcium carbonate extracted from the water.

Internal

Nervous System: The nervous system of an insect can be divided into a brain and a ventral nerve cord. The head capsule is made up of six fused segments, each with either a pair of ganglia, or a cluster of nerve cells outside of the brain. The first three pairs of ganglia are fused into the brain, while the three following pairs are fused into a structure of three pairs of ganglia under the insect's esophagus, called the subesophageal ganglion.

The thoracic segments have one ganglion on each side, which are connected into a pair, one pair per segment. This arrangement is also seen in the abdomen but only in the first eight segments. Many species of insects have reduced numbers of ganglia due to fusion or reduction. Some cockroaches have just six ganglia in the abdomen, whereas the wasp *Vespa crabro* has only two in the thorax and three in the abdomen. Some insects, like the house fly *Musca domestica*, have all the body ganglia fused into a single large thoracic ganglion.

At least a few insects have nociceptors, cells that detect and transmit sensations of pain. This was discovered in 2003 by studying the variation in reactions of larvae of the common fruitfly Drosophila to the touch of a heated probe and an unheated one. The larvae reacted to the touch of the heated probe with a stereotypical rolling behaviour that was not exhibited when the larvae were touched by the unheated probe. Although nociception has been demonstrated in insects, there is no consensus that insects feel pain consciously but see Pain in invertebrates.

Digestive System: An insect uses its digestive system to extract nutrients and other substances from the food it consumes. Most of this food is ingested in the form of macromolecules and other complex substances like proteins, polysaccharides, fats and nucleic acids. These macromolecules must be broken down by catabolic reactions into smaller molecules like amino acids and simple sugars before being used by cells of the body for energy, growth, or reproduction. This break-down process is known as digestion.

The main structure of an insect's digestive system is a long enclosed tube called the alimentary canal, which runs lengthwise through the body. The alimentary canal directs food unidirectionally from the mouth to the anus. It has three sections, each of which performs a different process of digestion. In addition to the alimentary canal, insects also have paired salivary glands and salivary reservoirs. These structures usually reside in the thorax, adjacent to the foregut.

The salivary glands (element 30 in numbered diagram) in an insect's mouth produce saliva. The salivary ducts lead from the glands to the reservoirs and then forward through the head to an opening called the salivarium, located behind the hypopharynx. By moving its mouthparts (element 32 in numbered diagram) the insect can mix its food with saliva. The mixture of saliva and food then travels through the salivary tubes into the mouth, where it begins to break down. Some insects, like flies, have extra-oral digestion. Insects using extra-oral digestion expel digestive enzymes onto their food to break it down. This strategy allows insects to extract a significant proportion of the available nutrients from the food source. The gut is where almost all of insects' digestion takes place. It can be divided into the foregut, midgut and hindgut.

Foregut

The first section of the alimentary canal is the foregut (element 27 in numbered diagram), or stomodaeum. The foregut is lined with a cuticular lining made of chitin and proteins as protection from tough food. The foregut includes the buccal cavity (mouth), pharynx, esophagus and crop and proventriculus (any part may be highly modified) which both store food and signify when to continue passing onward to the midgut.

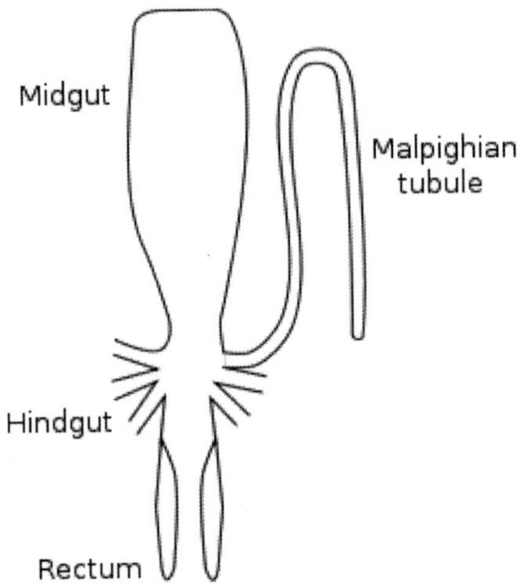

Figure: *Stylized diagram of insect digestive tract showing malpighian tubule, from an insect of the order Orthoptera.*

Digestion starts in buccal cavity (mouth) as partially chewed food is broken down by saliva from the salivary glands. As the salivary glands produce fluid and carbohydrate-digesting enzymes (mostly amylases), strong muscles in the pharynx pump fluid into the buccal cavity, lubricating the food like the salivarium does, and helping blood feeders, and xylem and phloem feeders. From there, the pharynx passes food to the esophagus, which could be just a simple tube passing it on to the crop and proventriculus, and then onward to the midgut, as in most insects. Alternately, the foregut may expand into a very enlarged crop and proventriculus, or the crop could just be a diverticulum, or fluid-filled structure, as in some Diptera species.

Midgut

Once food leaves the crop, it passes to the midgut (element 13 in numbered diagram), also known as the mesenteron, where the majority of digestion takes place. Microscopic projections from the midgut wall, called microvilli, increase the surface area of the wall and allow more nutrients to be absorbed; they tend to be close to the origin of the midgut. In some insects, the role of the microvilli and where they are located may vary. For example, specialised microvilli producing digestive enzymes may more likely be near the end of the midgut, and absorption near the origin or beginning of the midgut.

Hindgut

In the hindgut (element 16 in numbered diagram), or proctodaeum, undigested food particles are joined by uric acid to form fecal pellets. The rectum absorbs 90% of the water in these fecal pellets, and the dry pellet is then eliminated through the anus (element 17), completing the process of digestion. The uric acid is formed using hemolymph waste products diffused from the Malpighian tubules (element 20). It is then emptied directly into the alimentary canal, at the junction between the midgut and hindgut. The number of Malpighian tubules possessed by a given insect varies between species, ranging from only two tubules in some insects to over 100 tubules in others.

Edocrine System

The salivary glands (element 30 in numbered diagram) in an insect's mouth produce saliva. The salivary ducts lead from the glands to the reservoirs and then forward through the head to an opening called the salivarium, located behind the hypopharynx. By moving its mouthparts (element 32 in numbered diagram) the insect can mix its food with saliva. The mixture of saliva and food then travels through the salivary tubes into the mouth, where it begins to break down. Some insects, like flies, have extra-oral digestion. Insects using extra-oral digestion expel digestive enzymes onto their food to break it down. This strategy allows insects to extract a significant proportion of the available nutrients from the food source.

Reproductive System

The reproductive system of female insects consist of a pair of ovaries, accessory glands, one or more spermathecae, and ducts connecting these parts. The ovaries are made up of a number of egg tubes, called ovarioles, which vary in size and number by species. The number of eggs that the insect is able to make vary by the number of ovarioles with the rate that eggs can be develop being also influenced by ovariole design. Female insects are able make eggs, receive and store sperm, manipulate sperm from different males, and lay eggs. Accessory glands or glandular parts of the oviducts produce a variety of substances for sperm maintenance, transport and fertilization, as well as for protection of eggs. They can produce glue and protective substances for coating eggs or tough coverings for a batch of eggs called oothecae. Spermathecae are tubes or sacs in which sperm can be stored between the time of mating and the time an egg is fertilized.

For males, the reproductive system is the testis, suspended in the body cavity by tracheae and the fat body. Most male insects have a pair of testes, inside of which are sperm tubes or follicles that are enclosed within a membranous sac. The follicles connect to the vas deferens by the vas efferens, and the two tubular vasa deferentia connect to a median ejaculatory duct that leads to the outside. A portion of the vas deferens is often enlarged to form the seminal vesicle, which stores the sperm before they are discharged into the female. The seminal vesicles have glandular linings that secrete nutrients for nourishment and maintenance of the sperm. The ejaculatory duct is derived from an invagination of the epidermal cells during development and, as a result, has a cuticular lining. The terminal portion of the ejaculatory duct may be sclerotized to form the intromittent organ, the aedeagus. The remainder of the male reproductive system is derived from embryonic mesoderm, except for the germ cells, or spermatogonia, which descend from the primordial pole cells very early during embryogenesis.

Respiratory and Circulatory Systems

Insect respiration is accomplished without lungs. Instead, the insect respiratory system uses a system of internal tubes and sacs through which gases either diffuse or are actively pumped, delivering oxygen directly to tissues that need it via their trachea (element 8 in numbered diagram). Since oxygen is delivered directly, the circulatory system is not used to carry oxygen, and is therefore greatly reduced. The insect circulatory system has no veins or arteries, and instead consists of little more than a single, perforated dorsal tube which pulses peristaltically. Toward the thorax, the dorsal tube (element 14) divides into chambers and acts like the insect's heart. The opposite end of the dorsal tube is like the aorta of the insect circulating the hemolymph, arthropods' fluid analog of blood, inside the body cavity. Air is taken in through openings on the sides of the abdomen called spiracles.

The respiratory system is an important reason that limits the size of insects. As insects get bigger, this type of oxygen transport gets less efficient and thus the heaviest insect currently weighs less than 100 g. However, with increased atmospheric oxygen levels, as happened in the late Paleozoic, larger insects were possible, such as dragonflies with wingspans of more than two feet.

There are many different patterns of gas exchange demonstrated by different groups of insects. Gas exchange patterns in insects can range from continuous and diffusive ventilation, to discontinuous gas exchange. During continuous gas exchange, oxygen is taken in and carbon dioxide is released in a continuous cycle. In discontinuous gas exchange, however, the insect takes in oxygen while it is active and small amounts of carbon dioxide are released when the insect is at rest. Diffusive ventilation is simply a form of continuous gas exchange that occurs by diffusion rather than physically taking in the oxygen. Some species of insect that are submerged also have adaptations to aid in respiration. As larvae, many insects have gills that can extract oxygen dissolved in water, while others need to rise to the water surface to replenish air supplies which may be held or trapped in special structures.

Reproduction and Development

Figure: A pair of Simosyrphus grandicornis *hoverflies mating in flight.*

The majority of insects hatch from eggs. The fertilization and development takes place inside the egg, enclosed by a shell (chorion) that consists of maternal tissue. In contrast to eggs of other arthropods, most insect eggs are drought resistant. This is because inside the chorion two additional membranes develop from embryonic tissue, the amnion and the serosa. This serosa secretes a cuticle rich in chitin that protects the embryo against desiccation. In Schizophora however the serosa does not develop, but these flies lay their eggs in damp places, such as rotting matter.

Some species of insects, like the cockroach *Blaptica dubia*, as well as juvenile aphids and tsetse flies, are ovoviviparous. The eggs of ovoviviparous animals develop entirely inside the female, and then hatch immediately upon being laid. Some other species, such as those in the genus of cockroaches known as *Diploptera*, are viviparous, and thus gestate inside the mother and are born alive. Some insects, like parasitic wasps, show polyembryony, where a single fertilized egg divides into many and in some cases thousands of separate embryos. Insects may be *univoltine*, *bivoltine* or *multivoltine*, i.e. they may have one, two or many broods (generations) in a year.

Other developmental and reproductive variations include haplodiploidy, polymorphism, paedomorphosis or peramorphosis, sexual dimorphism, parthenogenesis and more rarely hermaphroditism. In haplodiploidy, which is a type of sex-determination system, the offspring's sex is determined by the number of sets of chromosomes an individual receives. This system is typical in bees and wasps. Polymorphism is where a species may have different *morphs* or *forms*, as in the oblong winged katydid, which has four different varieties: green, pink and yellow or tan. Some insects may retain phenotypes that are normally only seen in juveniles; this is called paedomorphosis. In peramorphosis, an opposite sort of phenomenon, insects take on previously unseen traits after they have matured into adults. Many insects display sexual dimorphism, in which males and females have notably different appearances, such as the moth *Orgyia recens* as an exemplar of sexual dimorphism in insects.

Figure: *The different forms of the male (top) and female (bottom) tussock moth* Orgyia recens *is an example of sexual dimorphism in insects.*

Some insects use parthenogenesis, a process in which the female can reproduce and give birth without having the eggs fertilized by a male. Many aphids undergo a form of parthenogenesis, called cyclical parthenogenesis, in which they alternate between one or many generations of asexual and sexual reproduction. In summer, aphids are generally female and parthenogenetic; in the autumn, males may be produced for sexual reproduction. Other insects produced by parthenogenesis are bees, wasps and ants, in which they spawn males. However, overall, most individuals are female, which are produced by fertilization. The males are haploid and the females are diploid. More rarely, some insects display hermaphroditism, in which a given individual has both male and female reproductive organs.

Insect life-histories show adaptations to withstand cold and dry conditions. Some temperate region insects are capable of activity during winter, while some others migrate to a warmer climate or go into a state of torpor. Still other insects have evolved mechanisms of diapause that allow eggs or pupae to survive these conditions.

Metamorphosis

Metamorphosis in insects is the biological process of development all insects must undergo. There are two forms of metamorphosis: incomplete metamorphosis and complete metamorphosis.

Incomplete Metamorphosis

Insects that show hemimetabolism, or incomplete metamorphosis, change gradually by undergoing a series of molts. An insect molts when it outgrows its exoskeleton, which does not stretch and would otherwise restrict the insect's growth. The molting process begins as the insect's epidermis secretes a new epicuticle. After this new epicuticle is secreted, the epidermis releases a mixture of enzymes that digests the endocuticle and thus detaches the old cuticle. When this stage is complete, the insect makes its body swell by taking in a large quantity of water or air, which makes the old cuticle split along predefined weaknesses where the old exocuticle was thinnest. Other arthropods have a much different process and only molt; though must accommodate for the difference in exoskeleton structure and make up with other enzymes.

Immature insects that go through incomplete metamorphosis are called nymphs or in the case of dragonflies and damselflies, naiads. Nymphs are similar in form to the adult except for the presence of wings, which are not developed until adulthood. With each molt, nymphs grow larger and become more similar in appearance to adult insects.

Complete Metamorphosis

Holometabolism, or complete metamorphosis, is where the insect changes in four stages, an egg or embryo, a larva, a pupa and the adult or imago. In these species, an egg hatches to produce a larva, which is generally worm-like in form. This worm-like form can be one of several varieties: eruciform (caterpillar-like), scarabaeiform (grub-like), campodeiform (elongated, flattened and active), elateriform (wireworm-like) or vermiform (maggot-like). The larva grows and eventually becomes a pupa, a stage marked by reduced movement and often sealed within a cocoon. There are three types of pupae: obtect, exarate or coarctate. Obtect pupae are compact, with the legs and other appendages enclosed. Exarate pupae have their legs and other appendages free and extended. Coarctate pupae develop inside the larval skin. Insects undergo considerable change in form during the pupal stage, and emerge as adults. Butterflies are a well-known example of insects that undergo complete metamorphosis, although most insects use this life cycle. Some insects have evolved this system to hypermetamorphosis.

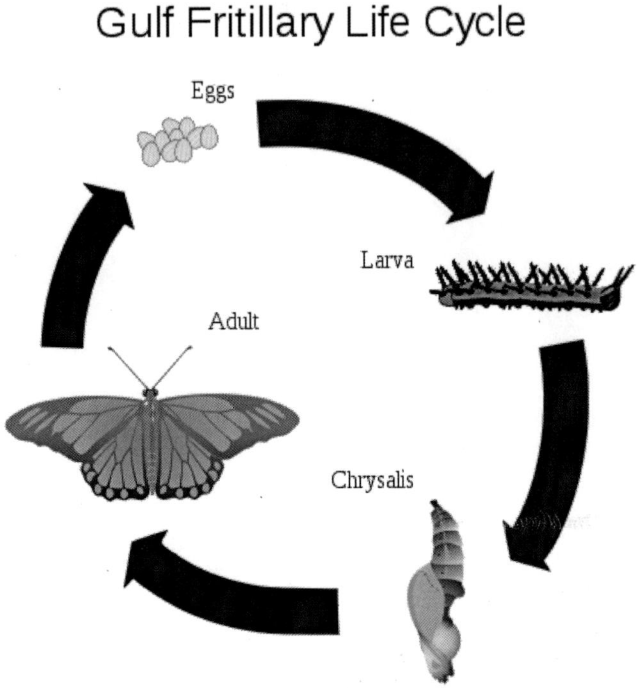

Figure: *Gulf Fritillary life cycle, an example of holometabolism.*

Some of the oldest and most successful insect groups, such Endopterygota, use a system of complete metamorphosis. Complete metamorphosis is unique to a group of certain insect orders including Diptera, Lepidoptera and Hymenoptera. This form of development is exclusive and not seen in any other arthropods.

Senses and Communication

Many insects possess very sensitive and, or specialised organs of perception. Some insects such as bees can perceive ultraviolet wavelengths, or detect polarized light, while the antennae of male moths can detect the pheromones of female moths over distances of many kilometres. There is a pronounced tendency for there to be a trade-off between visual acuity and chemical or tactile acuity, such that most insects with well-developed eyes have reduced or simple antennae, and vice versa. There are a variety of different mechanisms by which insects perceive sound, while the patterns are not universal, insects can generally hear sound if they can produce it. Different insect species can have varying hearing, though most insects can hear only a narrow range of frequencies related to the frequency of the sounds they can produce. Mosquitoes have been found to hear up to 2 kHz., and some grasshoppers can hear up to 50 kHz. Certain predatory and parasitic insects can detect the characteristic sounds made by their prey or hosts, respectively. For instance, some nocturnal moths can perceive the ultrasonic emissions of bats, which helps them avoid predation. Insects that feed on blood have special sensory structures that can detect infrared emissions, and use them to home in on their hosts.

Some insects display a rudimentary sense of numbers, such as the solitary wasps that prey upon a single species. The mother wasp lays her eggs in individual cells and provides each egg with a number of live caterpillars on which the young feed when hatched. Some species of wasp always provide five, others twelve, and others as high as twenty-four caterpillars per cell. The number of caterpillars is different among species, but always the same for each sex of larva. The male solitary wasp in the genus Eumenes is smaller than the female, so the mother of one species supplies him with only five caterpillars; the larger female receives ten caterpillars in her cell.

Light Production and Vision

Figure: *Insects have compound eyes and two antennae.*

A few insects, such as members of the families Poduridae and Onychiuridae (Collembola), Mycetophilidae (Diptera) and the beetle families Lampyridae, Phengodidae, Elateridae and Staphylinidae are bioluminescent. The most familiar group are the fireflies, beetles of the family Lampyridae. Some species are able to control this light generation to produce flashes. The function varies with some species using them to attract mates, while others use them to lure prey. Cave dwelling larvae of *Arachnocampa* (Mycetophilidae, Fungus gnats) glow to lure small flying insects into sticky strands of silk. Some fireflies of the genus *Photuris* mimic the flashing of female *Photinus* species to attract males of that species, which are then captured and devoured. The colours of emitted light vary from dull blue (*Orfelia fultoni*, Mycetophilidae) to the familiar greens and the rare reds (*Phrixothrix tiemanni*, Phengodidae).

Most insects, except some species of cave crickets, are able to perceive light and dark. Many species have acute vision capable of detecting minute movements. The eyes may include simple eyes or ocelli as well as compound eyes of varying sizes. Many species are able to detect light in the infrared, ultraviolet and the visible light wavelengths. Colour vision has been demonstrated in many species and phylogenetic analysis suggests that UV-green-blue trichromacy existed from at least the Devonian period between 416 and 359 million years ago.

Sound Production and Hearing

Insects were the earliest organisms to produce and sense sounds. Insects make sounds mostly by mechanical action of appendages. In grasshoppers and crickets, this is achieved by stridulation. Cicadas make the loudest sounds among the insects by producing and amplifying sounds with special modifications to their body and musculature. The African cicada *Brevisana brevis* has been measured at 106.7 decibels at a distance of 50 cm (20 in). Some insects, such as the *Helicoverpa zea*moths, hawk moths and Hedylid butterflies, can hear ultrasound and take evasive action when they sense that they have been detected by bats. Some moths produce ultrasonic clicks that were once thought to have a role in jamming bat echolocation. The ultrasonic clicks were subsequently found to be produced mostly by unpalatable moths to warn bats, just as warning colourations are used against predators that hunt by sight. Some otherwise palatable moths have evolved to mimic these calls.

Very low sounds are also produced in various species of Coleoptera, Hymenoptera, Lepidoptera, Mantodea and Neuroptera. These low sounds are simply the sounds made by the insect's movement. Through microscopic stridulatory structures located on the insect's muscles and joints, the normal sounds of the insect moving are amplified and can be used to warn or communicate with other insects. Most sound-making insects also have tympanal organs that can perceive airborne sounds. Some species in Hemiptera, such as the corixids (water boatmen), are known to communicate via underwater sounds. Most insects are also able to sense vibrations transmitted through surfaces.

Communication using surface-borne vibrational signals is more widespread among insects because of size constraints in producing air-borne sounds. Insects cannot effectively produce low-frequency sounds, and high-frequency sounds tend to disperse more in a dense environment (such as foliage), so insects living in such environments communicate primarily using substrate-borne vibrations. The mechanisms of production of vibrational signals are just as diverse as those for producing sound in insects.

Some species use vibrations for communicating within members of the same species, such as to attract mates as in the songs of the shield bug *Nezara viridula.* Vibrations can also be used to communicate between entirely different species; lycaenid (gossamer-winged butterfly) caterpillars which are myrmecophilous (living in a mutualistic association with ants) communicate with ants in this way. The Madagascar hissing cockroach has the ability to press air through its spiracles to make a hissing noise as a sign of aggression; the Death's-head Hawkmoth makes a squeaking noise by forcing air out of their pharynx when agitated, which may also reduce aggressive worker honey bee behaviour when the two are in close proximity.

Chemical Communication

Chemical communications in animals rely on a variety of aspects including taste and smell. Chemoreception is the physiological response of a sense organ (i.e. taste or smell) to a chemical stimulus where the chemicals act as signals to regulate the state or activity of a cell. A semiochemical is a message-carrying chemical that is meant to attract, repel, and convey information. Types of semiochemicals include pheromones and kairomones. One example is the butterfly *Phengaris arion* which uses chemical signals as a form of mimicry to aid in predation.

In addition to the use of sound for communication, a wide range of insects have evolved chemical means for communication. These chemicals, termed semiochemicals, are often derived from plant metabolites include those meant to attract, repel and provide other kinds of information. Pheromones, a type of semiochemical, are used for attracting mates of the opposite sex, for aggregating conspecific individuals of both sexes, for deterring other individuals from approaching, to mark a trail, and to trigger aggression in nearby individuals. Allomonea benefit their producer by the effect they have upon the receiver. Kairomones benefit their receiver instead of their producer. Synomones benefit the producer and the receiver. While some chemicals are targeted at individuals of the same species, others are used for communication across species. The use of scents is especially well known to have developed in social insects.

Social Behaviour

Social insects, such as termites, ants and many bees and wasps, are the most familiar species of eusocial animal. They live together in large well-organised colonies that may be so tightly integrated and genetically similar that the colonies of some species are sometimes considered superorganisms. It is sometimes argued that the various species of honey bee are the only invertebrates (and indeed one of the few non-human groups) to have evolved a system of abstract symbolic communication where a behaviour is used to *represent* and convey specific information about something in the environment. In this communication system, called dance language, the angle at which a bee dances represents a direction relative to the sun, and the length of the dance represents the distance to be flown.

Only insects which live in nests or colonies demonstrate any true capacity for fine-scale spatial orientation or homing. This can allow an insect to return unerringly to a single hole a few millimetres in diameter among thousands of apparently identical holes clustered together, after a trip of up to several kilometres' distance. In a phenomenon known as philopatry, insects that hibernate have shown the ability to recall a specific location up to a year after last viewing the area of interest. A few insects seasonally migrate large distances between different geographic regions (e.g., the overwintering areas of the Monarch butterfly).

Care of Young

The eusocial insects build nest, guard eggs, and provide food for offspring full-time. Most insects, however, lead short lives as adults, and rarely interact with one another except to mate or compete for mates. A small number exhibit some form of parental care, where they will at least guard their eggs, and sometimes continue guarding their offspring until adulthood, and possibly even feeding them. Another simple form of parental care is to construct a nest (a burrow or an actual construction, either of which may be simple or complex), store provisions in it, and lay an egg upon those provisions. The adult does not contact the growing offspring, but it nonetheless does provide food. This sort of care is typical for most species of bees and various types of wasps.

Locomotion

Flight

Insects are the only group of invertebrates to have developed flight. The evolution of insect wings has been a subject of debate. Some entomologists suggest that the wings are from paranotal lobes, or extensions from the insect's exoskeleton called the nota, called the *paranotal theory*. Other theories are based on a pleural origin. These theories include suggestions that wings originated from modified gills, spiracular flaps or as from an appendage of the epicoxa. The *epicoxal theory* suggests the insect wings are modified epicoxal exites, a modified appendage at the base of the legs or coxa. In the Carboniferous age, some of the *Meganeura* dragonflies had as much as a 50 cm (20 in) wide wingspan. The appearance of gigantic insects has been found to be consistent with high atmospheric oxygen. The respiratory system of insects constrains their size, however the high oxygen in the atmosphere allowed larger sizes. The largest flying insects today are much smaller and include several moth species such as the Atlas moth and the White Witch (*Thysania agrippina*). Insect flight has been a topic of great interest in aerodynamics due partly to the inability of steady-state theories to explain the lift generated by the tiny wings of insects.

Unlike birds, many small insects are swept along by the prevailing winds although many of the larger insects are known to make migrations. Aphids are known to be transported long distances by low-level jet streams. As such, fine line patterns associated with converging winds within weather radar imagery, like the WSR-88D radar network, often represent large groups of insects.

Walking

Many adult insects use six legs for walking and have adopted a tripedal gait. The tripedal gait allows for rapid walking while always having a stable stance and has been studied extensively in cockroaches. The legs are used in alternate triangles touching the ground. For the first step, the middle right leg and the front and rear left legs are in contact with the ground and move the insect forward, while the front and rear right leg and the middle left leg are lifted and moved forward to a new position. When they touch the ground to form a new stable triangle the other legs can be lifted and brought forward in turn and so on. The purest form of the tripedal gait is seen in insects moving at high speeds. However, this type of locomotion is not rigid and insects can adapt a variety of gaits. For example, when moving slowly, turning, or avoiding obstacles, four or more feet may be touching the ground.

Insects can also adapt their gait to cope with the loss of one or more limbs. Cockroaches are among the fastest insect runners and, at full speed, adopt a bipedal run to reach a high velocity in proportion to their body size. As cockroaches move very quickly, they need to be video recorded at several hundred frames per second to reveal their gait. More sedate locomotion is seen in the stick insects or walking sticks (Phasmatodea). A few insects have evolved to walk on the surface of the water, especially members of the Gerridae family, commonly known as water striders. A few species of ocean-skaters in the genus *Halobates* even live on the surface of open oceans, a habitat that has few insect species.

Use in Robotics

Insect walking is of particular interest as an alternative form of locomotion in robots. The study of insects and bipeds has a significant impact on possible robotic methods of transport. This may allow new robots to be designed that can traverse terrain that robots with wheels may be unable to handle.

Swimming

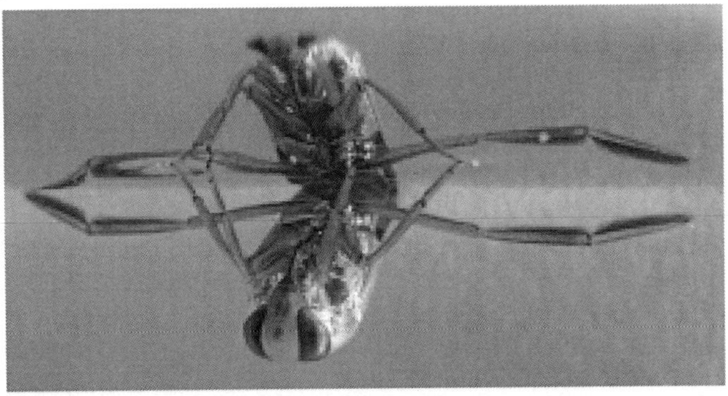

Figure: *The backswimmer* Notonecta glauca *underwater, showing its paddle-like hindleg adaptation.*

A large number of insects live either part or the whole of their lives underwater. In many of the more primitive orders of insect, the immature stages are spent in an aquatic environment. Some groups of insects, like certain water beetles, have aquatic adults as well.

Many of these species have adaptations to help in under-water locomotion. Water beetles and water bugs have legs adapted into paddle-like structures. Dragonfly naiads use jet propulsion, forcibly expelling water out of their rectal chamber. Some species like the water striders are capable of walking on the surface of water. They can do this because their claws are not at the tips of the legs as in most insects, but recessed in a special groove further up the leg; this prevents the claws from piercing the water's surface film. Other insects such as the Rove beetle *Stenus* are known to emit pygidial gland secretions that reduce surface tension making it possible for them to move on the surface of water by Marangoni propulsion (also known by the German term *Entspannungsschwimmen*).

Ecology

Insect ecology is the scientific study of how insects, individually or as a community, interact with the surrounding environment or ecosystem. Insects play one of the most important roles in their ecosystems, which includes many roles, such as soil turning and aeration, dung burial, pest control, pollination and wildlife nutrition. An example is the beetles, which are scavengers that feed on dead animals and fallen trees and thereby recycle biological materials into forms found useful by other organisms. These insects, and others, are responsible for much of the process by which topsoil is created.

Defence and Predation

Figure: *Perhaps one of the most well-known examples of mimicry, the viceroy butterfly (top) appears very similar to the noxious-tasting monarch butterfly (bottom).*

Insects are mostly soft bodied, fragile and almost defenceless compared to other, larger lifeforms. The immature stages are small, move slowly or are immobile, and so all stages are exposed to predation and parasitism. Insects then have a variety of defence strategies to avoid being attacked by predators or parasitoids. These include camouflage, mimicry, toxicity and active defence.

Camouflage is an important defence strategy, which involves the use of colouration or shape to blend into the surrounding environment. This sort of protective colouration is common and widespread among beetle families, especially those that feed on wood or vegetation, such as many of the leaf beetles (family Chrysomelidae) or weevils. In some of these species, sculpturing or various coloured scales or hairs cause the beetle to resemble bird dung or other inedible objects. Many of those that live in sandy environments blend in with the colouration of the substrate. Most phasmids are known for effectively replicating the forms of sticks and leaves, and the bodies of some species (such as *O. macklotti* and *Palophus centaurus*) are covered in mossy or lichenous outgrowths that supplement their disguise. Some species have the ability to change colour as their surroundings shift (*B. scabrinota*, *T. californica*). In a further behavioural adaptation to supplement crypsis, a number of species have been noted to perform a rocking motion where the body is swayed from side to side that is thought to reflect the movement of leaves or twigs swaying in the breeze. Another method by which stick insects avoid predation and resemble twigs is by feigning death (catalepsy), where the insect enters a motionless state that can be maintained for a long period. The nocturnal feeding habits of adults also aids Phasmatodea in remaining concealed from predators.

Another defence that often uses colour or shape to deceive potential enemies is mimicry. A number of longhorn beetles (family Cerambycidae) bear a striking resemblance to wasps, which helps them avoid predation even though the beetles are in fact harmless. Batesian and Müllerian mimicry complexes are commonly found in Lepidoptera. Genetic polymorphism and natural selection give rise to otherwise edible species (the mimic) gaining a survival advantage by resembling inedible species (the model). Such a mimicry complex is referred to as *Batesian* and is most commonly known by the mimicry by the limenitidine Viceroy butterfly of the inedible danaine Monarch. Later research has discovered that the Viceroy is, in fact more toxic than the Monarch and this resemblance should be considered as a case of Müllerian mimicry. In Müllerian mimicry, inedible species, usually within a taxonomic order, find it advantageous to resemble each other so as to reduce the sampling rate by predators who need to learn about the insects' inedibility. Taxa from the toxic genus *Heliconius* form one of the most well known Müllerian complexes.

Chemical defence is another important defence found amongst species of Coleoptera and Lepidoptera, usually being advertised by bright colours, such as the Monarch butterfly. They obtain their toxicity by sequestering the chemicals from the plants they eat into their own tissues. Some Lepidoptera manufacture their own toxins. Predators that eat poisonous butterflies and moths may become sick and vomit violently, learning not to eat those types of species; this is actually the basis of Müllerian mimicry. A predator who has previously eaten a poisonous lepidopteran may avoid other species with similar markings in the future, thus saving many other species as well. Some ground beetles of the Carabidae family can spray chemicals from their abdomen with great accuracy, to repel predators.

Pollination

Figure: *European honey bee carrying pollen in a pollen basket back to the hive*

Pollination is the process by which pollen is transferred in the reproduction of plants, thereby enabling fertilisation and sexual reproduction. Most flowering plants require an animal to do the transportation. While other animals are included as pollinators, the majority of pollination is done by insects. Because insects usually receive benefit for the pollination in the form of energy rich nectar it is a grand example of mutualism. The various flower traits (and combinations thereof) that differentially attract one type of pollinator or another are known as pollination syndromes. These arose through complex plant-animal adaptations. Pollinators find flowers through bright colourations, including ultraviolet, and attractant pheromones. The study of pollination by insects is known as *anthecology*.

Parasitism

Many insect are parasites of other insects such as the parasitoid wasps. These insects are known as entomophagous parasites. They can be beneficial due to their devastation of pests that can destroy crops and other resources. Many insects have a parasitic relationship with humans such as the mosquito. These insects are known to spread diseases such as malaria and yellow fever and because of such, mosquitoes indirectly cause more deaths of humans than any other animal.

12

Integrated Pest Management Philosophy

The philosophy of integrated pest control is based upon (1) the principles of applied ecology, (2) the recognition that 10000 mortality of all pests is not required, and (3)avoidance of disruptive chemicals. Since this philosophy of pest control is more of a system of principles for guidance than a set of rules in a fixed program, it requires more than the usual judgment to make it work on the farm. Integrated pest control is based on the principles of applied ecology. Whatever is done in any field situation is always founded, as far as possible, upon full knowledge of the ecology of the pests. In practice, this is attainable only by continuous sampling of the insects in all parts of the fields where integrated control programs are being applied. By this policy of watching the fields much more closely than usual, all the natural control factors can be exploited. Implementing this policy requires the services of a professional insect ecologist under contract to the farmer. This specialist in integrated pest control is the leader of a team that samples the fields, analyzing the total interacting complex of organisms, including the pests and their biological controls, and the effects of all of the agricultural practices and industrial activities of man. This dynamic consideration of the fields takes into account the evolving patterns of plant growth, seasonal developments, and the ever–changing populations of organisms associated with them. Emphasis is always on the developing insect population levels and the predictions as to the course that the pest populations will take.

The supervising insect ecologist implements successful pest management in the fields by applying any or all practical cultural and biological practices known to suppress the pests. To attain a favourable balance in a field, it is necessary to attract, augment, and conserve all possible numbers of beneficial predators, parasites, and disease organisms in the fields. Even low numbers of insectary-grown and/or field harvested beneficial insects can be colonized.

Best results from integrated pest control are obtained when effective natural enemies of most target pests are present in the complex of organisms. Crop rotations to alfalfa, grains, and pasture, along with planned cover crops, bring greater plant diversity to a farming area. This diversity of plants brings greater diversity in both host and prey species, and provides nectar, pollen, and alternate host organisms as food for predators and parasites. Alfalfa, grains, and pasture grasses are normally under such satisfactory biological control that they usually require very little control other than cultural management. Such management programs in alfalfa, using strip harvesting techniques, aids in preventing sudden emigration of pests at cutting time. With less disruptive action to pests and their biological controls, there is greater stability in population interaction, and less chance for one species to dominate and cause damage. Different natural enemies have different efficiency levels related to pest population densities; the above environmental manipulations help provide a complete complex of predators and parasites, at a very low cost.

Field harvesting techniques can selectively remove portions of pest populations and collect many predators, parasites, and disease organisms for use in augmenting the natural controls in other fields. Releases of this material, along with colonizations of insectary-grown beneficial insects, can help bring the field into a favourable balance in time to prevent economic losses. Every pest spot in a field that is controlled biologically adds to the total number of predators in the field; every biologically controlled field grows more predators for the nearby fields and the agricultural area. The less you spray, the less you need to spray. This complex of beneficial insects is a natural resource that the farmer must protect and conserve by applying pesticides only when absolutely necessary to protect the crop. For certain pests there are no immediate prospects for development of effective pest suppression techniques. Then pest populations increase beyond tolerable limits and there is no predictable chance of obtaining a favourable balance, the crop can always be sprayed.

This leads us to the second principle of integrated pest control; a recognition that 100% mortality of all pests is not required to prevent economic losses. This may seem elementary, but society has been brainwashed into believing that "the only good bug is a dead bug"; it is important to point out that insects, mites, weeds, etc. are pests only when they affect our way of life beyond tolerable limits, It is essential to the concept of integrated control that we live with a few pests in order to obtain and retain a favourable balance.

Strict standardization procedures that do not necessarily give the consumer food of better quality or taste may force the farmer to spray for light infestations of a pest. Often any plant damage that might occur would not be harmful to the crop, except perhaps in appearance. Plants can tolerate low populations of aphids, thrips, and mites with no intolerable damage affecting profitable production of the crop. Instead of spraying, the insect ecologist uses these minor pests as food in colonizing additional insectary-grown or field harvested predators. The challenge for the future is to devise ways to grow and or field harvest sufficient beneficial insects and apply them to the fields by mechanical means; in this way, biological control could be better implemented and guaranteed, and many more fields could be saved from pesticides.

The third principle of integrated pest control is the avoidance of disruptive chemicals. Not all pesticides have adverse effects, and certain dosages of conventional pesticides are less disruptive to the existing biological control. Backup sprays are applied when necessary, but of a kind and in a manner least disruptive. Integrated pest control seeks to suppress target pest populations below tolerable damage levels without destroying their biological controls. The professional insect ecologist must know how to use pesticides in this way in given situations. However, because of the polluting side effects of too much pesticide in the environment, it is even more important that he know when and where not to use them.

There is not time to dwell on all the pest control problems that have developed as a result of complete reliance on chemicals; a short list would include target pest resistance to pesticides, target pest resurgence following treatments, secondary pest outbreaks, pesticide residue problems in the crop, the soil, or drift to nearby areas, or runoff to streams and drainage, and destruction of pollinators and other forms of life beneficial to man. In many instances these side effects have created havoc, and the undesirable results have more than off-set any original benefits. It is becoming more and more difficult to kill the target pests economically.

One often-underestimated side problem is the drift of pesticide residues onto nearby crop areas and the widespread effect of destroying favourable biological balances. Searching predators and parasites moving over the plants in search of prey are exposed to more residue, and are killed in greater numbers than the more sedentary pests. This uncontrollable drift creates new pest situations in nearby fields, that in turn require treatment: the problem is spread further and further. Such drift effects have triggered pest outbreaks over large areas and destroyed many beneficial insects. The more you spray the more you need to spray.

The cost of chemical pest control programs is high when all of these side effects begin to interact. The crops may be saved, but at great cost to society and the farmer (who is stuck with the pesticide bills). The real measure of success in farming is in the profits; the real problem that causes the farmer to look for new answers to pest control is the spiraling cost of repeated and ineffective treatments. This search for new answers has forced farmers and entomologists to modify their pest control practices, so as to avoid disruptive chemicals as much as possible.

How effective is integrated pest control in practice? Experience has shown that this philosophy, when practiced on the farm by competent professional insect ecologist, is the only pest management system dynamic enough to solve the problems and keep up with the changing agricultural problems of the future. The farmer must willingly support a team effort to carry out practical application of integrated pest control. The leader is the insect ecologist, a highly trained specialist who utilises materials and techniques as complicated as those used by the pharmacist or veterinarian. New, more sophisticated integrated pest control programs of the future will be implemented by these professional insect ecologists. The backup part of the integrated pest control team is the insectary, which grows the predators and parasites, and field harvests other beneficial insects, for use in applying positive biological and environmental pest suppression. Other members of the team are the researchers, specialising in long-term ecological studies and development of integrated pest control techniques. Laboratory research specialists on physiology and behaviour can discover much to aid in interpretation of data, but only studies in the fields will reveal the true interplay of all the complex ecological factors. The usual fragmentary short-term experiments, so common in pesticide evaluations, are more than vulnerable to error, when considered in the long-term ecological sense.

It is important to understand that valid scientific discovery can result from careful observations and recordings of recurring natural happenings. In other words, scientific proof can be obtained from a sequence of experiments or experiences. History will repeat itself, provided the same combination of effective conditions recurs. In applying integrated pest control, the insect ecologist frequently samples and assesses the fields in his care, constituting a continuous experiment in applied ecology not duplicated at research level. This experiment is carried out in many different field situations, throughout the season, and often over many years on the same farm. What better proof is there of the success of integrated pest control than the practical experience of profitable production of quality products?

Research specialists in integrated pest control must work closely with the insect ecologist, who is the foreman in charge of pest control. Since a multifactored environment is not easy to understand, the researcher can aid in interpreting the experience of the applied worker. At the same time, the insect ecologist can provide assistance to the researchers economically, gathering samples, etc., because he lives in the area of the field experiment. In this way the practical limitations of time and money can be somewhat overcome, and the work of solving the pressing problem of pest control can continue. The commercial pest control applicators and the producers of chemicals and other pest control products will continue to make an important contribution. The pesticide companies will make new products to answer the needs of integrated pest control, because it is the more selective target pesticides that are least disruptive that the farmers will buy.

Active and aggressive educational programs supporting the ecological approach to pest control can help separate fact from fiction regarding integrated pest management. General recommendations to farmers should be flexible, to allow time for alternative integrated pest control techniques to be exploited. Adjustable economic injury levels, including the assessment of beneficial organisms and suggestions for use of selective pesticides, are advisable.

The combined efforts of all members of the integrated pest control team are needed to insure successful and profitable production of sufficient food and fibre for the future. Successful pest control is relative, but in the final analysis it is measured in the economic sense. High costs of ineffective pesticide programs and their ultimate effects on food and fibre prices concerns farmer and consumer alike. There is no question of the success of integrated control in the minds of farmers who have changed their philosophy of pest control and are "suddenly" not paying excessive pesticide bills. The first year or two of successful integrated pest control is often attributed to "luck" or a "good year" by critics of such programs. After several years of economical pest management with little or no spray bills, the farmer eagerly support their insect ecologist and the insectary that supports him. As food supplies become shorter in the coming decades, and as consumers become more conscious of high food costs and toxic residues in foods and less conscious of appearance only, the artificial standards that pesticides depend on will be relaxed. More sophisticated and safer processing of farm products will be developed to insure removal of insect parts, with the result that the consumer will have bug-free as well as toxicant-free products.

Insect Management

A successful insect management program can best be accomplished by combining IPM techniques, such as accurate pest identification, scouting, monitoring and action thresholds, with biological and alternative (preventative) control practices and selective insecticide applications.

Use scouting and monitoring techniques to help assess and quantify insect populations over time.

Insecticides should be used only when action thresholds have been exceeded or damaging insect populations are present. Action thresholds help minimize insecticide use and crop damage. Avoid making routine insecticide applications to crops without evidence of insect activity and damage. All insects have natural enemies which, if conserved, can help regulate pest populations. Whenever possible, use selective insecticides that spare beneficial organisms and target the pest you wish to control. Broad-spectrum insecticides should be used as a last resort. All insecticide applications should be made with ample water and with nozzles directed so that they provide thorough coverage of the plant parts where insects hide. Alternate between insecticide classes or families to help manage insect resistance and extend the life of available products.

Insect Identification

Become familiar with the biology and life-cycle of the major insect pests that attack crops on an annual basis. Understanding some basic insect biology often reveals when the pest is most vulnerable to control measures and helps lead to successful management efforts. Insects usually have either a simple life-cycle, where they grow from egg to nymph to adult, or a slightly more complex cycle, where they mature from egg to larva and then go through metamorphosis in the pupal stage before becoming adults. Insects damage plants either as nymphs/ larvae or as adults, or in both immature and adult stages. Learn to recognise the important life-stages of major insect pests, and to recognise the crop injury or damage they produce to help determine when control efforts are needed. You should also be able to identify common beneficial insects and their immature stages (i.e., lady bug larvae) to avoid accidentally targeting natural enemies with insecticides. There are dozens of minor and secondary insect pests that may attack crops on a less frequent basis. Growers should have unknown insects and minor pests identified when they are suspected of causing crop damage. Misidentification often leads to the application of an ineffective pesticide and extensive or chronic crop damage. The application of ineffective or unnecessary pesticides can often reduce the populations of beneficial organisms or natural enemies and lead to secondary insect or mite outbreaks. An accurate diagnosis of the problem early on can often avoid years of frustration and needless expense.

Scouting and Monitoring

Crops should be inspected or scouted for insects or damage in a systematic fashion, on a regular basis throughout the growing season. For many crops and pests, this may mean walking fields weekly, or even more frequently, especially during critical or vulnerable plant development stages. Crops should be scouted in a systematic fashion by walking in an "M" or "W" pattern as you crisscross the field. Select plants (i.e., 25 or 50) at random and quantify the pest damage or count the individual insects. Record the average number of insects or damage per plant for each field. Scouting crops always saves money in the long run by allowing for early pest detection, by reducing crop damage and by helping to maintain consistent quality. Sometimes scouting duties can be performed by other farm members or by hired consultants.

Monitoring insect populations with various types of traps (black light, pheromone, sticky, baited) can supplement or sometimes substitute for information normally gathered during crop scouting. Insect traps can help you quantify pest pressure that is difficult to see, such as the number of night flying (i.e., corn earworm) moths present. The number of insects captured in traps is often used to time scouting activities, predict future pest levels, or is used in conjunction with action thresholds to time sprays and help avoid crop damage.

Action Thresholds

Action thresholds tell you when to spray to prevent economic damage to the crop. They also help you avoid applying insecticides to crops when insect populations are low or no pests are present. Thresholds can be based on the number of insects found per plant, the amount of injury or damage per plant, or the number of insects captured in a trap. Using action thresholds helps improve insecticide timing and effectiveness, helps minimize the number of applications and associated costs, and helps reduce crop damage. Use action thresholds whenever possible to help you decide if and when a spray is needed. Some insect thresholds are provided in this publication.

Preventative Controls

As mentioned in the introduction to the general Pest Management section, there are many preventative management options, such as cultural, mechanical, physical, genetic and biological controls. These should be used whenever possible to help prevent insect pest problems. Preventative insect control options can be used alone or in combination to provide a complete management program. For example, plant inspections, eliminating weeds in the greenhouse, using screens, avoiding excess nitrogen applications, using plastic mulch and preserving natural enemies can combine to produce a very effective aphid management program. If the distance between fields is great enough, crop rotation (alone) for Colorado potato beetle and cucumber beetles can often keep these tough pests from reaching damaging levels for the entire season. However, if only nearby fields are available, CPB can be effectively controlled by the use of a combination of two or more alternative controls, such as short-distance crop rotation, intervening trap crops or plastic-lined trenches, row covers, organic mulches, flaming, and microbial controls. Many preventative controls are mentioned in crop-specific IPM manuals, individual pest fact sheets and on IPM web sites.

Perimeter Trap Cropping utilises a combination of control measures to concentrate and/or kill pests in the border area of the field. This technique involves planting one or more rows of an attractive plant species so that it completely encircles the less attractive main crop, and intercepts a migrating insect population. Often a chemical or microbial insecticide is used to spot-spray the perimeter trap crop (only), sparing beneficials on the main crop, and substantially reducing insecticide use. Perimeter trap cropping has been shown to be effective at controlling diamondback moth larvae on cabbage, using collards as the trap crop; pepper maggots on bell peppers or eggplant, using hot cherry peppers as the trap crop; and cucumber beetles on summer squash, winter squash, cucumbers, melons, and sometimes pumpkins, using Blue Hubbard squash or another Cucurbita maxima variety as the trap crop. This technique may prove useful on related crops and against additional pests in the future.

Insecticides

Follow the label specifications for application rates. If the rate suggested in this guide does not agree with the current label, follow the label recommendations. Amounts of pesticides are in lb/A (pounds per acre), oz/A (ounces per acre) or pt/A (pints per acre) of commercial formulation, not a.i. (active ingredient), unless otherwise stated. The percentage by weight or concentration of a.i. varies with different brands of pesticides. The label gives pounds of a.i. in each commercial preparation. For example, Lorsban 4E has 4 lb active ingredient/gallon of formulated material.

The amount of insecticide recommended per acre should not be changed when varying gallonage of water per acre. Make sure a pesticide is labelled for concentrate application before using a low volume sprayer, air or mist sprayer, aircraft or other concentrate application equipment.

Seed Treatments

Insecticide seed treatments may help protect your crop from soil insects and other early-season pests. Some of the seed treatments listed under various crops must be applied by a professional licensed seed coating applicator. In some cases the use or purchase of treated seed may be economical as it may result in a substantial reduction in pesticide use and increase the plant stand and yield in the field.

Resistance Management

Adult pests that survive an environmental hardship, such as the application of an insecticide, are likely to pass the trait that enabled them to survive on to their offspring. Repeat applications with the same type of pesticide will eventually remove almost all the susceptible individuals from a pest population and leave only those with the resistant gene. Because insects and mites go through generations quickly, resistance to an overused pesticide can develop in as little as three years.

To preserve the useful life of a pesticide, it is necessary to practice resistance management. The most effective way to extend the useful life of an effective product is to use it on a single pest generation only, and then on the next generation use a second pesticide with a different method of killing the pest (mode of action). On some pests particularly prone to developing resistance (i.e., Colorado potato beetle) it is best to use a product with a particular mode of action on one generation every second year only. To help select pesticides with a different mode of action.

Resistance Groups (number and letter) indicate products with a common mode of action. For multiple applications to the same crop, select products from different resistance groups. Once a pest develops resistance to a group of pesticides with a particular mode of action, a higher rate of a similar chemical from the same group usually will not control the insect.

There are many other techniques that can help delay the onset of resistance. Most resistance management techniques help minimize the use of pesticides, so that a lower proportion of each pest generation is exposed to the toxin.

- Integrate chemical control with effective cultural, mechanical, physical, and genetic management options.
- Use biological/microbial control agents or other selective pesticides to preserve natural enemies and help minimize the number of repeat applications.
- Scout, monitor and use action thresholds to ensure that applications are necessary.
- Good spray coverage helps do the job right the first time and avoids unnecessary repeat applications: use the proper size nozzles and the correct angle or orientation and an adequate amount of water per acre.
- Time the application so that the most vulnerable insect life stage is exposed to the spray.
- Use spot sprays, perimeter trap crop treatments, refuge plantings, and other methods that prevent the entire field or population from being treated to help preserve susceptible individuals.

Toxicity Rating of Insecticides

Insecticides vary greatly in their toxicity to humans and the environment. The toxicity of the insecticide is usually stated in the precaution on the label. For example, a skull and crossbones figure and the signal word "Danger" are always found on the label of highly toxic (Toxicity Class I) materials. Those of medium toxicity (Toxicity Class II) carry the signal word "Warning." The least toxic materials (Toxicity Class III) have the signal word "Caution." The toxicity of a pesticide is expressed in terms of oral and dermal LD50. LD50 is the dosage of poison that kills 50% of test animals (usually rats or rabbits) with a single application of the pure pesticide for a given weight of the animal (mg/kg of body weight). The lower the LD_{50} value, the more toxic the material. Oral LD50 is the measure of the toxicity of pure pesticide when administered internally to test animals. Dermal LD50 is the measure of the toxicity of pure pesticide applied to the skin of test animals. Generally, an oral exposure is more toxic than a dermal one.

Vegetable Pest Identification

In nature, there are always some garden pests chewing on plants; that's just the way it is. However, not all pest damage is significant enough to warrant action. Even the healthiest gardens encounter bugs at one time or another, yet they still produce a beautiful harvest. As gardeners, we must each consider the level of pest activity that we are willing to tolerate. Every garden is susceptible to a pest invasion - tucking into foliage, feasting on the vegetable patch and burrowing through the bedding.

The best way to maintain a healthy garden is to educate yourself and learn to identify common "bad bugs." Inspect your garden regularly to detect problems early. The sooner a pest is identified the easier it will be to manage using earth-friendly methods.

Ants

Ants are a common pest in the garden, especially on the lawn, sandy ground, plants and fruit trees. They are prolific when it comes to creating nests and usually do so in a sunny area of the garden. They actually do little serious damage to plants, although they can be a nuisance and weaken both garden and greenhouse plants.

How to Treat Ants

Have a dust around plants and along runs with a good ant killer powder or spray, which are widely available. For chemical-free ant control, homemade pest repellents provide the best form of prevention, and there are certain smells they detest. Try out mint, camphor and clove oil. You can also grow your own ant repellent in the vegetable garden, by mixing hot peppers in a blender with a bit of water to create a thick mash, which you can spread in problem areas. Use a stiff brush to remove ant hills in the lawn.

Black and Greenfly (Aphids)

Greenfly or blackfly - also known as aphids - are a major plant pest, sucking the sap of many plants. These small green or brown insects gather on shoot tips, flower buds and the undersides of young leaves, and are very common on all types of plants. They can also deposit a sticky honeydew on which disease spores can stick, and leaves may twist or yellow.

How to Treat Aphids

Tackle aphids as soon as they are spotted. Spray with a greenfly killer insecticide or, for an organic option, as aphids breathe through their skin, spraying with a diluted washing up liquid will clog their skin and cause them to suffocate quickly. For plants which are affected outdoors, try to encourage birds with seeds and tables near plants at risk from aphids as they will feast away.

Bulb Fly (Daffodil & Narcissus Bulb Fly)

This small black and brown fly swoops just as the foliage is dying down on the bulbs, after flowering, and lays eggs in the crown of the bulb. The eggs hatch and become the maggot, which bores down into the bulb to feed on the inside. Daffodil and narcissus bulbs become soft, squidgy and rotten when the inside is eaten out by the fly's maggot, and very few leaves or flowers will be produced.

How to Treat Bulb Fly

Eliminate bulb flies with a strong mixture of insecticides and water, or for organic control, apply diatomaceous earth, available at garden centres, to help control freshly hatched grubs. Consider covering over the planting after bloom with a lightweight garden cover, such as fine netting.

Planthoppers

A planthopper is any insect in the infraorder Fulgoromorpha within the Hemiptera. The name comes from their remarkable resemblance to leaves and other plants of their environment and from the fact that they often "hop" for quick transportation in a similar way to that of grasshoppers. However, planthoppers generally walk very slowly so as not to attract attention. Distributed worldwide, all members of this group are plant-feeders, though surprisingly few are considered pests. The infraorder contains only a single superfamily, Fulgoroidea. Fulgoroids are most reliably distinguished from the other members of the classical "Homoptera" by two features; the bifurcate ("Y"-shaped) anal vein in the forewing, and the thickened, three-segmented antennae, with a generally round or egg-shaped second segment (pedicel) that bears a fine filamentous arista.

Nymphs of many Fulgoroids produce wax from special glands on the abdominal terga and other parts of the body. These are hydrophobic and help conceal the insects. Adult females of many families also produce wax which may be used to protect eggs. Planthoppers are often vectors for plant diseases, especially phytoplasmas which live in the phloem of plants and can be transmitted by planthoppers when feeding.

Cixiidae

The Cixiidae are a family of fulgoroid insects, one of many families commonly known as planthoppers, distributed worldwide and comprising more than 2,000 species from over 150 genera. The genera are placed into three subfamilies, Borystheninae, Bothriocerinae and Cixiinae with sixteen tribes currently accepted in Cixiinae.

Cixiid species are typically comparatively small (body size less than a centimetre) and usually inconspicuous. Nymphs live underground, feeding on roots. Adults feed on herbs, shrubs and/or trees; some are polyphagous, while others are specialised on their host plants (monophagous). A couple of species are cavernicolous, feeding on roots in volcanic caves. Females occasionally bear impressive "wax tails" produced by wax-producing plates at the tip of their abdomen.

Several species are of economic importance (e.g. *Hyalesthes obsoletus*, *Haplaxius crudus*). Phytoplasma are common parasites in these insects, causing diseases in coconut palms and foliage, grapevines, sugar beets, and lilies.

The fossil record of Cixiidae is limited, and a number of taxa which have placed into the family may need to be reexamined and moved to different families. The oldest confirmed taxa are from the Early Cretaceous with *'Cixius' petrinus* described from Barremian deposits in England, *Karebodopoides aptianus* from Hauterivian to Aptian Lebanese amber and Cretofennahia cretacea plus an unnamed specimen from the Aptian of Brazil.

Due to the abundant nature of Cixiidae as inclusions in Eocene Baltic amber a number of taxa have been described, including *Glisachaemus jonasdamzeni* and *Autrimpus sambiorum*. Several taxa have also been described from Miocene Dominican amber including *Oligocixia electrina* and *Oliarius kulickae*.

Many gardeners are misidentifying the white insects on plant stems. In most cases it is planthoppers rather than mealybugs. The planthoppers are not destructive and jump away when touched. Mealybugs stay put. A water blast is all that's needed for the planthoppers. The insect problems are interesting and diverse this season. For mealybugs, I would probably try the EcoSMART insect control product. The line is in quite a few of the retail stores now. Another option is to spray the Mound Drench mixture. It has worked for me in the past. Mix equal parts orange oil, compost tea and liquid molasses. Use 4 oz per gallon of water for the spray formula. It can also be made by adding 2 oz of orange oil to the Garrett Juice formula. This method gives the plants foliar feeding as well as insect control.

COMMON NAME: Leafhopper, Planthopper SCIENTIFIC NAME: Order *Homoptera*, family Cicadellidae, many species SIZE: Adult—1/10" to 1/2" IDENTIFICATION: Adults are distinctive wedge-shaped insects with triangular heads. Most are brown or green, but some are brightly decorated. All have well-developed hind legs for jumping. Nymphs are very similar to adults, but paler in colour and wingless. BIOLOGY AND LIFE CYCLE: Incomplete metamorphosis. Adults overwinter and lay eggs in rows or clusters when leaves emerge. Eggs hatch in ten to fourteen days, nymphs develop for one to four weeks. Two to five generations a year. HABITAT: Fruits, vegetables, and ornamental plants. FEEDING HABITS: Adults and nymphs suck juices from stems and undersides of leaves, causing a mottled look. ECONOMIC IMPORTANCE: Toxic saliva can cause stunty leaf growth or warty, crinkled, or rolled edges. Some are vectors of plant disease. NATURAL CONTROL: Maintain healthy plants. Protect natural enemies such as parasitic wasps and flies, damsel bugs, minute pirate bugs, lady beetles, lacewings, and spiders. Birds, frogs, lizards, and insectivorous insects. ORGANIC CONTROL: Strong blasts of water and citrus oil products. Horticultural or plant oil for heavy infestations. INSIGHT: Entomologists say that leafhoppers are responsible for spreading more than 150 plant disorders.

Beetle

The Coleoptera order of insects is commonly called beetles. The word "coleoptera" is from the Greek êïëåüò, *koleos*, meaning "sheath"; and ðôåñüí, *pteron*, meaning "wing", thus "sheathed wing", because most beetles have two pairs of wings, the front pair, the "elytra", being hardened and thickened into a sheath-like, or shell-like, protection for the rear pair, and for the rear part of the beetle's body. The superficial consistency of most beetles' morphology, in particular their possession of elytra, has long suggested that the Coleoptera are monophyletic, but growing evidence indicates this is unjustified, there being arguments, for example, in favour of allocating the current suborder Adephaga their own order, or very likely even more than one.

The Coleoptera include more species than any other order, constituting almost 25% of all known types of animal life-forms. About 40% of all described insect species are beetles (about 400,000 species), and new species are discovered frequently. Some estimates put the total number of species, described and undescribed, at as high as 100 million, but a figure of one million is more widely accepted. The largest taxonomic family is commonly thought to be the Curculionidae (the weevils or snout beetles), but recently the Staphylinidae (the rove beetles) have claimed this title.

The diversity of beetles is very wide. They are found in all major habitats, except marine and the polar regions. They have many classes of ecological effects; particular species are adapted to practically every kind of diet. Some are nonspecialist detritus feeders, breaking down animal and plant debris; some feed on particular kinds of carrion such as flesh or hide; some feed on wastes such as dung; some feed on fungi, some on particular species of plants, others on a wide range of plants. Some are generalist pollen, flower and fruit eaters. Some are predatory, usually on other invertebrates; some are parasites or parasitoids. Many of the predatory species are important controls of agricultural pests. For example, beetles in the family Coccinellidae ("ladybirds" or "ladybugs") consume aphids, scale insects, thrips, and other plant-sucking insects that damage crops.

Conversely, beetles are prey of various invertebrates and vertebrates, including insects, fish, reptiles, birds, and mammals. The Coleoptera are not generally serious pests, but they include agricultural and industrial pests, such as the Colorado potato beetle *Leptinotarsa decemlineata*, the boll weevil *Anthonomus grandis*, the red flour beetle *Tribolium castaneum*, and the mungbean or cowpea beetle *Callosobruchus maculatus*. Also included is the death-watch beetle, the larvae of which can cause serious structural damage to buildings by boring into the timbers.

Species in the Coleoptera have a hard exoskeleton, particularly on their forewings (elytra, singular elytron). These elytra distinguish beetles from most other insect species, except for the Dermaptera. The hemelytra of Heteroptera have a slight resemblance, but are not the same and their function is largely different.

Like all armoured insects, beetles' exoskeletons comprise numerous plates called sclerites, some fused, and some separated by thin sutures. This combines armored defences with maintaining flexibility. The general anatomy of a beetle is superficially uniform, but specific organs and appendages may vary greatly in appearance and function between the many families in the order, and even more so between the suborders (such as Adephaga) that currently seem increasingly to be separate orders in their own right. All insects' bodies are divided into three sections: the head, the thorax, and the abdomen, and the Coleoptera are no exception. Their internal morphology and physiology also resemble those of other insects.

Beetles are endopterygotes; they undergo complete metamorphosis, a biological process by which an animal physically develops after a birth or hatching, undergoing a series of conspicuous and relatively abrupt changes in its body structure. Males may fight for females in various ways, and such species tend to display marked sexual dimorphism.

Distribution and Diversity

Figure: *Coccinellidaes in County Wexford, Ireland.*

Beetles are by far the largest order of insects, with 350,000–400,000 species in four suborders (Adephaga, Archostemata, Myxophaga, and Polyphaga), making up about 40% of all insect species described, and about 30% of all animals. Though classification at the family level is a bit unstable, about 500 families and subfamilies are recognised. One of the first proposed estimates of the total number of beetle species on the planet is based on field data rather than on catalog numbers. The technique used for this original estimate, possibly as many as 12 million species, was criticized, and was later revised, with estimates of 850,000–4,000,000 species proposed. Some 70–95% of all beetle species, depending on the estimate, remain undescribed. The beetle fauna is not equally well known in all parts of the world. For example, the known beetle diversity of Australia is estimated at 23,000 species in 3265 genera and 121 families. This is slightly lower than reported for North America, a land mass of similar size with 25,160 species in 3526 genera and 129 families. While other predictions show there could be as many as 28,000 species in North America, including those currently undescribed, a realistic estimate of the little-studied Australian beetle fauna's true diversity could vary from 80,000 to 100,000.

Coleoptera are found in nearly all natural habitats, including freshwater and marine habitats, everywhere vegetative foliage is found, from trees and their bark to flowers, leaves, and underground near roots- even inside plants in galls, in every plant tissue, including dead or decaying ones.

External Morphology

Beetles are generally characterized by a particularly hard exoskeleton and hard forewings (elytra). The beetle's exoskeleton is made up of numerous plates, called sclerites, separated by thin sutures. This design provides armored defences while maintaining flexibility. The general anatomy of a beetle is quite uniform, although specific organs and appendages may vary greatly in appearance and function between the many families in the order. Like all insects, beetles' bodies are divided into three sections: the head, the thorax, and the abdomen.

Head

The head, having mouthparts projecting forward or sometimes downturned, is usually heavily sclerotized and varies in size. The eyes are compound and may display remarkable adaptability, as in the case of whirligig beetles (family Gyrinidae), where they are split to allow a view both above and below the waterline. Other species also have divided eyes – some longhorn beetles (family Cerambycidae) and weevils

– while many have eyes that are notched to some degree. A few beetle genera also possess ocelli, which are small, simple eyes usually situated farther back on the head (on the vertex).

Figure: Scarabaeus viettei *(syn.* Madateuchus viettei, *Scarabaeidae) showing a "shovel head" adaptation*

Figure: *Head of* Cephalota circumdata, *showing the compound eyes and mouthparts*

Beetles' antennae are primarily organs of smell, but may also be used to feel a beetle's environment physically. They may also be used in some families during mating, or among a few beetles for defence. Antennae vary greatly in form within the Coleoptera, but are often similar within any given family. In some cases, males and females of the same species will have different antennal forms. Antennae may be clavate (flabellate and lamellate are subforms of clavate, or clubbed antennae), filiform, geniculate, moniliform, pectinate, or serrate.

Beetles have mouthparts similar to those of grasshoppers. Of these parts, the most commonly known are probably the mandibles, which appear as large pincers on the front of some beetles. The mandibles are a pair of hard, often tooth-like structures that move horizontally to grasp, crush, or cut food or enemies. Two pairs of finger-like appendages, the maxillary and labial palpi, are found around the mouth in most beetles, serving to move food into the mouth. In many species, the mandibles are sexually dimorphic, with the males' enlarged enormously compared with those of females of the same species.

Thorax

The thorax is segmented into the two discernible parts, the pro- and pterathorax. The pterathorax is the fused meso- and metathorax, which are commonly separate in other insect species, although flexibly articulate from the prothorax. When viewed from below, the thorax is that part from which all three pairs of legs and both pairs of wings arise. The abdomen is everything posterior to the thorax. When viewed from above, most beetles appear to have three clear sections, but this is deceptive: on the beetle's upper surface, the middle "section" is a hard plate called the pronotum, which is only the front part of the thorax; the back part of the thorax is concealed by the beetle's wings. This further segmentation is usually best seen on the abdomen.

Figure: Acilius sulcatus, *a diving beetle showing hind legs adapted for life in water*

Extremities

The multisegmented legs end in two to five small segments called tarsi. Like many other insect orders, beetles bear claws, usually one pair, on the end of the last tarsal segment of each leg. While most beetles use their legs for walking, legs may be variously modified and adapted for other uses. Among aquatic families – Dytiscidae, Haliplidae, many species of Hydrophilidae and others – the legs, most notably the last pair, are modified for swimming and often bear rows of long hairs to aid this purpose. Other beetles have fossorial legs that are widened and often spined for digging. Species with such adaptations are found among the scarabs, ground beetles, and clown beetles (family Histeridae). The hind legs of some beetles, such as flea beetles (within Chrysomelidae) and flea weevils (within Curculionidae), are enlarged and designed for jumping.

Wings

The elytra are connected to the pterathorax, so named because it is where the wings are connected (*pteron* meaning "wing" in Greek). The elytra are not used for flight, but tend to cover the hind part of the body and protect the second pair of wings (*alae*). They must be raised to move the hind flight wings. A beetle's flight wings are crossed with veins and are folded after landing, often along these veins, and stored below the elytra. A fold (*jugum*) of the membrane at the base of each wing is a characteristic feature. In some beetles, the ability to fly has been lost. These include some ground beetles (family Carabidae) and some "true weevils" (family Curculionidae), but also desert- and cave-dwelling species of other families. Many have the two elytra fused together, forming a solid shield over the abdomen. In a few families, both the ability to fly and the elytra have been lost, with the best known example being the glow-worms of the family Phengodidae, in which the females are larviform throughout their lives.

Abdomen

The abdomen is the section behind the metathorax, made up of a series of rings, each with a hole for breathing and respiration, called a spiracle, composing three different segmented sclerites: the tergum, pleura, and the sternum. The tergum in almost all species is membranous, or usually soft and concealed by the wings and elytra when not in flight. The pleura are usually small or hidden in some species, with each pleuron having a single spiracle. The sternum is the most widely visible part of the abdomen, being a more or less scelortized segment. The abdomen itself does not have any appendages, but some (for example, Mordellidae have articulating sternal lobes.

Internal Morphology

Digestive System

The digestive system of beetles is primarily based on plants, upon which they, for the most part, feed, with mostly the anterior midgut performing digestion, although in predatory species (for example Carabidae), most digestion occurs in the crop by means of midgut enzymes. In Elateridae species, the predatory larvae defecate enzymes on their prey, with digestion being extraorally. The alimentary canal basically consists of a short, narrow pharynx, a widened expansion, the crop, and a poorly developed gizzard. After is the midgut, that varies in dimensions between species, with a large amount of cecum, with a hindgut, with varying lengths. Typically, four to six Malpighian tubules occur.

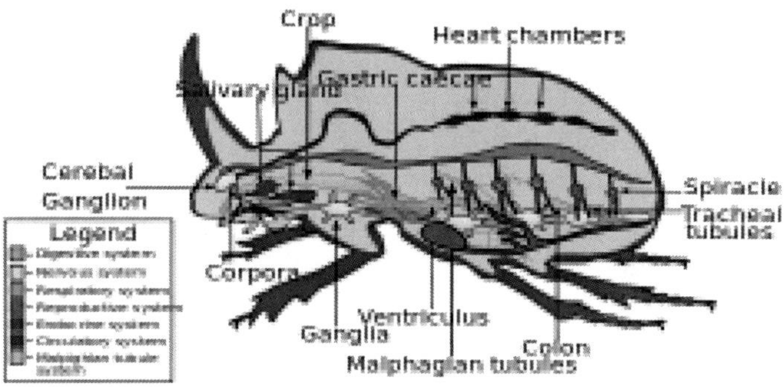

Figure: *A diagram showing the general internal anatomy of beetles*

Nervous System

The nervous system in beetles contains all the types found in insects, varying between different species, from three thoracic and seven or eight abdominal ganglia which can be distinguished to that in which all the thoracic and abdominal ganglia are fused to form a composite structure.

Respiratory System

Like most insects, beetles inhale oxygen and exhale carbon dioxide via a tracheal system. Air enters the body through spiracles, and circulates within the haemocoel in a system of tracheae and tracheoles, through the walls of which the relevant gases can diffuse appropriately.

Diving beetles, such as the Dytiscidae, carry a bubble of air with them when they dive. Such a bubble may be contained under the elytra or against the body by specialised hydrophobic hairs. The bubble covers at least some of the spiracles, thereby permitting the oxygen to enter the tracheae.

The function of the bubble is not so much as to contain a store of air, to act as a physical gill. The air that it traps is in contact with oxygenated water, so as the animal's consumption depletes the oxygen in the bubble, more oxygen can diffuse in to replenish it. Carbon dioxide is more soluble in water than either oxygen or nitrogen, so it readily diffuses out faster than in. Nitrogen is the most plentiful gas in the bubble, and the least soluble, so it constitutes a relatively static component of the bubble and acts as a stable medium for respiratory gases to accumulate in and pass through. Occasional visits to the surface are sufficient for the beetle to re-establish the constitution of the bubble.

Circulatory System

Like other insects, beetles have open circulatory systems, based on hemolymph rather than blood. Also as in other insects, a segmented tube-like heart is attached to the dorsal wall of the hemocoel. It has paired inlets or *ostia* at intervals down its length, and circulates the hemolymph from the main cavity of the haemocoel and out through the anterior cavity in the head.

Specialised Organs

Different glands specialise for different pheromones produced for finding mates. Pheromones from species of Rutelinea are produced from epithelial cells lining the inner surface of the apical abdominal segments; amino acid-based pheromones of Melolonthinae are produced from eversible glands on the abdominal apex. Other species produce different types of pheromones. Dermestids produce esters, and species of Elateridae produce fatty acid-derived aldehydes and acetates. For means of finding a mate also, fireflies (Lampyridae) use modified fat body cells with transparent surfaces backed with reflective uric acid crystals to biosynthetically produce light, or bioluminescence. The light produce is highly efficient, as it is produced by oxidation of luciferin by enzymes (luciferases) in the presence of adenosine triphosphate (ATP) and oxygen, producing oxyluciferin, carbon dioxide, and light.

A notable number of species have developed special glands to produce chemicals for deterring predators. The ground beetle's (of Carabidae) defencive glands, located at the posterior, produce a variety of hydrocarbons, aldehydes, phenols, quinones, esters, and acids released from an opening at the end of the abdomen. African carabid beetles (for example, *Anthia* and *Thermophilum* – Thermophilum generally included within *Anthia*) employ the same chemicals as ants: formic acid. Bombardier beetles have well-developed, like other carabid beetles, pygidial glands that empty from the lateral edges of the intersegment membranes between the seventh and eighth abdominal segments. The gland is made of two containing chambers. The first holds hydroquinones and hydrogen peroxide, with the second holding just hydrogen peroxide plus catalases. These chemicals mix and result in an explosive ejection, forming temperatures of around 100 °C (212 °F), with the breakdown of hydroquinone to $H_2 + O_2$ + quinone, with the O_2 propelling the excretion.

Tympanal organs or hearing organs, which is a membrane (tympanum) stretched across a frame backed by an air sac and associated sensory neurons, are described in two families. Several species of the genus *Cicindela* (Cicindelidae) have ears on the dorsal surfaces of their first abdominal segments beneath the wings; two tribes in the subfamily Dynastinae (Scarabaeidae) have ears just beneath their pronotal shields or neck membranes. The ears of both families are sensitive to ultrasonic frequencies, with strong evidence indicating they function to detect the presence of bats by their ultrasonic echolocation. Though beetles constitute a large order and live in a variety of niches, examples of hearing are surprisingly lacking amongst species, though likely most simply remain undiscovered.

Reproduction and Development

Beetles are members of the superorder Endopterygota, and accordingly most of them undergo complete metamorphosis. The typical form of metamorphosis in beetles passes through four main stages: the egg, the larva, the pupa, and the imago or adult. The larvae are commonly called grubs and the pupa sometimes is called the chrysalis. In some species, the pupa may be enclosed in a cocoon constructed by the larva towards the end of its final instar. Going beyond "complete metamorphosis", however, some beetles, such as typical members of the families Meloidae and Rhipiphoridae, undergo hypermetamorphosis in which the first instar takes the form of a triungulin.

Mating

Figure: *Punctate flower chafers (*Neorrhina punctata, *Scarabaeidae) mating*

Beetles may display extremely intricate behaviour when mating. Pheromone communication is likely to be important in the location of a mate. Different species use different chemicals for their pheromones. Some scarab beetles (for example, Rutelinae) utilise pheromones derived from fatty acid synthesis, while other scarab beetles use amino acids and terpenoid compounds (for example, Melolonthinae). Another way species of Coleoptera find mates is the use of biosynthesized light, or bioluminescence. This special form of a mating call is confined to fireflies (Lampyridae) by the use of abdominal light-producing organs. The males and females engage in complex dialogue before mating, identifying different species by differences in duration, flight patterns, composition, and intensity.

Before mating, males and females may engage in various forms of behaviour. They may stridulate, or vibrate the objects they are on. In some species (for example, Meloidae), the male climbs onto the dorsum of the female and strokes his antennae on her head, palps, and antennae. In the genus *Eupompha* of said family, the male draws the antennae along his longitudinal vertex. They may not mate at all if they do not perform the precopulatory ritual.

Conflict can play a part in the mating rituals of species such as burying beetles (genus *Nicrophorus*), where conflicts between males and females rage until only one of each is left, thus ensuring reproduction by the strongest and fittest. Many male beetles are territorial and will fiercely defend their small patches of territory from intruding males. In such species, the males may often have horns on their heads and/ or thoraces, making their overall body lengths greater than those of the females, unlike most insects. Pairing is generally quick, but in some cases will last for several hours. During pairing, sperm cells are transferred to the female to fertilize the egg.

Lifecycle

Egg: A single female may lay from several dozen to several thousand eggs during her lifetime. Eggs are usually laid according to the substrate on which the larvae will feed upon hatching. Among others, they can be laid loose in the substrate (for example, flour beetle), laid in clumps on leaves (for example, Colorado potato beetle), individually attached (for example, mungbean beetle and other seed borers), or buried in the medium (for example, carrot weevil).

Parental care varies between species, ranging from the simple laying of eggs under a leaf to certain scarab beetles, which construct underground structures complete with a supply of dung to house and feed their young.

Other beetles are leaf rollers, biting sections of leaves to cause them to curl inwards, then laying their eggs, thus protected, inside.

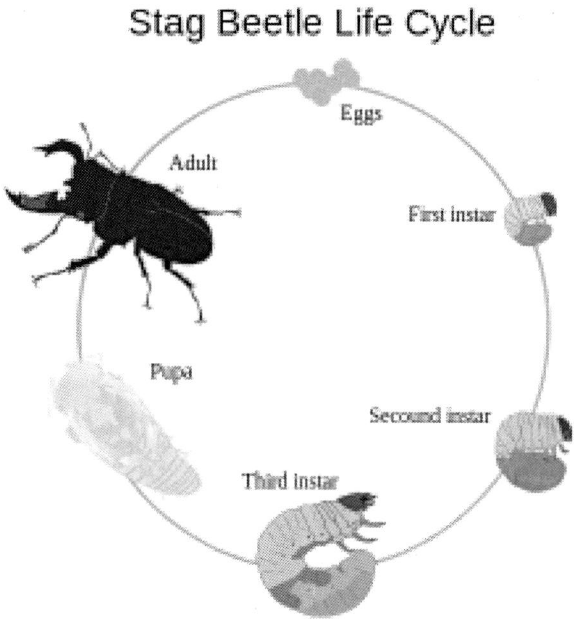

Figure: *The lifecycle of the stag beetle includes three instars.*

Larva: The larva is usually the principal feeding stage of the beetle lifecycle. Larvae tend to feed voraciously once they emerge from their eggs. Some feed externally on plants, such as those of certain leaf beetles, while others feed within their food sources. Examples of internal feeders are most Buprestidae and longhorn beetles. The larvae of many beetle families are predatory like the adults (ground beetles, ladybirds, rove beetles). The larval period varies between species, but can be as long as several years. The larvae are highly varied amongst species, with well-developed and sclerotized heads, and have distinguishable thoracic and abdominal segments (usually the tenth, though sometimes the eighth or ninth).

Beetle larvae can be differentiated from other insect larvae by their hardened, often darkened heads, the presence of chewing mouthparts, and spiracles along the sides of their bodies. Like adult beetles, the larvae are varied in appearance, particularly between beetle families. Beetles whose larvae are somewhat flattened and are highly mobile are the ground beetles, some rove beetles, and others; their larvae are described as campodeiform. Some beetle larvae resemble hardened worms with dark head capsules and minute legs. These are elateriform larvae, and are found in the click beetle (Elateridae) and darkling beetle

(Tenebrionidae) families. Some elateriform larvae of click beetles are known as wireworms. Beetles in the families of the Scarabaeoidea have short, thick larvae described as scarabaeiform, but more commonly known as grubs.

Figure: *A scarabaeiform larva known as a curl grub*

All beetle larvae go through several instars, which are the developmental stages between each moult. In many species, the larvae simply increase in size with each successive instar as more food is consumed. In some cases, however, more dramatic changes occur. Among certain beetle families or genera, particularly those that exhibit parasitic lifestyles, the first instar (the planidium) is highly mobile to search out a host, while the following instars are more sedentary and remain on or within their host. This is known as hypermetamorphosis; examples include the blister beetles (family Meloidae) and some rove beetles, particularly those of the genus *Aleochara*.

Pupa: As with all endopterygotes, beetle larvae pupate, and from these pupae emerge fully formed, sexually mature adult beetles, or imagos. Adults have extremely variable lifespans, from weeks to years, depending on the species. In some species, the pupa may go through all four forms during its development, called hypermetamorphosis (for example, Meloidae). Pupae always have no mandibles (are adecticous). In most, the appendages are not attached to the pupae, or they are exarate; with most being obtect in form.

Behaviour

Locomotion

Aquatic beetles use several techniques for retaining air beneath the water's surface. Beetles of the family Dytiscidae hold air between the abdomen and the elytra when diving. Hydrophilidae have hairs on their under surface that retain a layer of air against their bodies. Adult crawling water beetles use both their elytra and their hind coxae (the basal segment of the back legs) in air retention, while whirligig beetles simply carry an air bubble down with them whenever they dive.

The elytra allow beetles and weevils to both fly and move through confined spaces, doing so by folding the delicate wings under the elytra while not flying, and folding their wings out just before take off. The unfolding and folding of the wings is operated by muscles attached to the wing base; as long as the tension on the radial and cubital veins remains, the wings remain straight. In day-flying species (for example, Buprestidae, Scarabaeidae), flight does not include large amounts of lifting of the elytra, having the metathorac wings extended under the lateral elytra margins.

Figure: *Photinus pyralis, firefly, in flight*

Communication

Beetles have a variety of ways to communicate, some of which include a sophisticated chemical language through the use of pheromones. From the host tree, mountain pine beetles have many forms of communication. They can emit both an aggregative pheromone and an anti-aggregative pheromone. The aggregative pheromone attracts other beetles to the tree, and the anti-aggregative pheromone neutralizes the aggregative pheromone. This helps to avoid the harmful effects of having too many beetles on one tree competing for resources. The mountain pine beetle can also stridulate to communicate, or rub body parts together to create sound, having a "scraper" on their abdomens that they rub against a grooved surface on the underside of their left wing cover to create a sound that is not audible to humans. Once the female beetles have arrived on a suitable pine tree host, they begin to stridulate and produce aggregative pheromones to attract other unmated males and females. New females arrive and do the same as they land and bore into the tree.

As the males arrive, they enter the galleries that the females have tunneled, and begin to stridulate to let the females know they have arrived, and to also warn others that the female in that gallery is taken. At this point, the female stops producing aggregative pheromones and starts producing anti-aggregative pheromone to deter more beetles from coming.

Since species of Coleoptera use environmental stimuli to communicate, they are affected by the climate. Microclimates, such as wind or temperature, can disturb the use of pheromones; wind would blow the pheromones while they travel through the air. Stridulating can be interrupted when the stimulus is vibrated by something else.

Parental Care

Figure: *A dung beetle rolling dung*

Among insects, parental care is very uncommon, only found in a few species. Some beetles also display this unique social behaviour. One theory states parental care is necessary for the survival of the larvae, protecting them from adverse environmental conditions and predators. One species, a rover beetle (*Bledius spectabilis*) displays both causes for parental care: physical and biotic environmental factors. Said species lives in salt marshes, so the eggs and/or larvae are endangered by the rising tide. The maternal beetle will patrol the eggs and larvae and apply the appropriate burrowing behaviour to keep them from flooding and from asphyxiating. Another advantage is that the mother protects the eggs and larvae from the predatory carabid beetle *Dicheirotrichus gustavi* and from the parasitoid wasp *Barycnemis blediator*. Up to 15% of larvae are killed by this parasitoid wasp, being only protected by maternal beetles in their dens.

Some species of dung beetle also display a form of parental care. Dung beetles collect animal feces, or "dung", from which their name is derived, and roll it into a ball, sometimes being up to 50 times their own weight; albeit sometimes it is also used to store food. Usually it is the male that rolls the ball, with the female hitch-hiking or simply following behind. In some cases the male and the female roll together. When a spot with soft soil is found, they stop and bury the dung ball. They will then mate underground. After the mating, one or both of them will prepare the brooding ball. When the ball is finished, the female lays eggs inside it, a form of mass provisioning. Some species do not leave after this stage, but remain to safeguard their offspring.

Feeding

Besides being abundant and varied, beetles are able to exploit the wide diversity of food sources available in their many habitats. Some are omnivores, eating both plants and animals. Other beetles are highly specialised in their diet. Many species of leaf beetles, longhorn beetles, and weevils are very host-specific, feeding on only a single species of plant. Ground beetles and rove beetles (family Staphylinidae), among others, are primarily carnivorous and will catch and consume many other arthropods and small prey, such as earthworms and snails. While most predatory beetles are generalists, a few species have more specific prey requirements or preferences.

Figure: Mylabris pustulata *(Meloidae) feeding on the petals of* Ipomoea carnea

Decaying organic matter is a primary diet for many species. This can range from dung, which is consumed by coprophagous species (such as certain scarab beetles of the family Scarabaeidae), to dead animals, which are eaten by necrophagous species (such as the carrion beetles of the family Silphidae). Some of the beetles found within dung and carrion are in fact predatory. These include the clown beetles, preying on the larvae of coprophagous and necrophagous insects.

Ecology

Defence and Predation

Figure: *Beetles may be preyed upon by other insects such as robber flies*

Beetles and their larvae have a variety of strategies to avoid being attacked by predators or parasitoids. These include camouflage, mimicry, toxicity, and active defence. Camouflage involves the use of colouration or shape to blend into the surrounding environment. This sort of protective colouration is common and widespread among beetle families, especially those that feed on wood or vegetation, such as many of the leaf beetles (family Chrysomelidae) or weevils. In some of these species, sculpturing or various coloured scales or hairs cause the beetle to resemble bird dung or other inedible objects.

Many of those that live in sandy environments blend in with the colouration of the substrate. The giant African longhorn beetle (*Petrognatha gigas*) resembles the moss and bark of the tree it feeds on. Another defence that often uses colour or shape to deceive potential enemies is mimicry.

A number of longhorn beetles (family Cerambycidae) bear a striking resemblance to wasps, which helps them avoid predation even though the beetles are in fact harmless. This defence is an example of Batesian mimicry and, together with other forms of mimicry and camouflage occurs widely in other beetle families, such as the Scarabaeidae. Beetles may combine their colour mimicry with behavioural mimicry, acting like the wasps they already closely resemble. Many beetle species, including ladybirds, blister beetles, and lycid beetles can secrete distasteful or toxic substances to make them unpalatable or even poisonous. These same species often exhibit aposematism, where bright or contrasting colour patterns warn away potential predators, and there are, not surprisingly, a great many beetles and other insects that mimic these chemically protected species.

Figure: Clytus arietis *(Cerambycidae), a wasp mimic*

Chemical defence is another important defence found amongst species of Coleoptera, usually being advertised by bright colours. Others may utilise behaviours that would be done when releasing noxious chemicals (for example, Tenebrionidae). Chemical defence may serve purposes other than just protection from vertebrates, such as protection from a wide range of microbes, and repellents.

Some species release chemicals in the form of a spray with surprising accuracy, such as ground beetles (Carabidae), may spray chemicals from their abdomen to repel predators. Some species take advantage of the plants from which they feed, and sequester the chemicals from the plant that would protect it and incorporate into their own defence. African carabid beetles (for example, *Anthia* and *Thermophilum*) employ the same chemicals used by ants, while Bombardier beetles have a their own unique separate gland, spraying potential predators from far distances.

Large ground beetles and longhorn beetles may defend themselves using strong mandibles and/or spines or horns to forcibly persuade a predator to seek out easier prey. Many species have large protrusions from their thorax and head such as the Rhinoceros beetle, which can be used to defended themselves from predators. Many species of weevil that feed out in the open on leaves of plants react to attack by employing a "drop-off reflex".

Even further, some will combine it with thanatosis, which they will close up their legs, antennae, mandibles, etc. and use their cryptic colouration to blend in with the background. Species with varied colouration do not do this as they can not camaflouge.

Parasitism

Over 1000 species of beetles are known to be either parasitic, predatory, or commensals in the nests of ants. Technically, many species of beetles and their larvae could be considered to be ectoparasites, because they feed on plants and some live inside the bark or wood of trees, but such relationships are generally regarded as herbivory (plant eating) rather than parasitism.

A few species of beetles are actually ectoparasitic on mammals. One such species, *Platypsyllus castoris*, parasitises beavers (*Castor* spp.). This beetle lives as a parasite both as a larva and as an adult, feeding on epidermal tissue and possibly on skin secretions and wound exudates. They are strikingly flattened dorsoventrally, no doubt as an adaptation for slipping between the beavers' hairs. They also are wingless and eyeless, as are many other ectoparasites.

Other parasitic beetles include those that are kleptoparasites of other invertebrates, such as the small hive beetle (*Aethina tumida*) that infests honey bee hives. The larvae tunnel through comb towards stored honey or pollen, damaging or destroying cappings and comb in the process. Larvae defecate in honey and the honey becomes discoloured from the feces, which causes fermentation and a frothiness in the honey; the honey develops a characteristic odor of decaying oranges. Damage and fermentation cause honey to run out of combs, destroying large amounts of it both in hives and sometimes also in honey extracting rooms. Heavy infestations cause bees to abscond; though the beetle is only a minor pest in Africa, beekeepers in other regions have reported the rapid collapse of even strong colonies.

Pollination

Beetle-pollinated flowers are usually large, greenish or off-white in colour, and heavily scented. Scents may be spicy, fruity, or similar to decaying organic material. Most beetle-pollinated flowers are flattened or dish-shaped, with pollen easily accessible, although they may include traps to keep the beetle longer. The plants' ovaries are usually well protected from the biting mouthparts of their pollinators. Beetles may be particularly important in some parts of the world such as semiarid areas of southern Africa and southern California and the montane grasslands of KwaZulu-Natal in South Africa.

Mutualism

Amongst most orders of insects, mutualism is not common, but some examples occur in species of Coleoptera, such as the ambrosia beetle, the ambrosia fungus, and probably bacteria.

The beetles excavate tunnels in dead trees in which they cultivate fungal gardens, their sole source of nutrition. After landing on a suitable tree, an ambrosia beetle excavates a tunnel in which it releases spores of its fungal symbiont.

The fungus penetrates the plant's xylem tissue, digests it, and concentrates the nutrients on and near the surface of the beetle gallery, so the weevils and the fungus both benefit. The beetles cannot eat the wood due to toxins, and uses its relationship with fungi to help overcome its host tree defences and to provide nutrition for their larvae.

Chemically mediated by a bacterially produced polyunsaturated peroxide, this mutualistic relationship between the beetle and the fungus is coevolved.

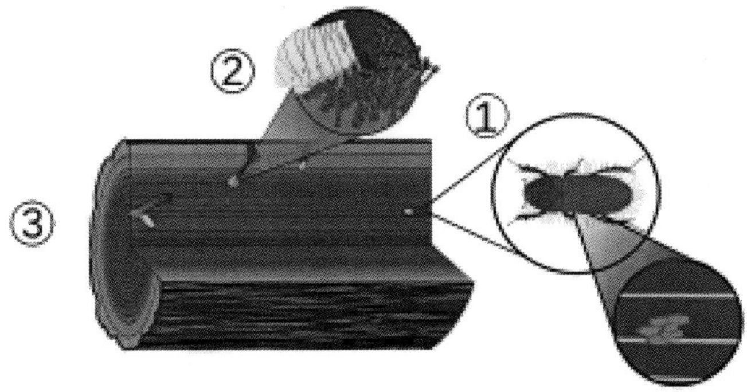

Figure: *1: The adult beetle burrows hole into wood and lays eggs, carrying fungal spores in its mycangia. 2: The larva feeds on the fungus, which digest the wood, removing toxins: they mutually benefit. 3: The larva pupates and then ecloses.*

Commensalism

Pseudoscorpions are small arachnids with flat, pear-shaped bodies and pincers that resemble those of scorpions (only distant relatives), usually ranging from 2 to 8 millimetres (0.08 to 0.31 in) in length. Their small size allows them to hitch rides under the elytra of giant harlequin beetles to be dispersed over wide areas while simultaneously being protected from predators. They may also find mating partners as other individuals join them on the beetle. This would be a form of parasitism if the beetle were harmed in the process, but the beetle is, presumably, unaffected by the presence of the hitchhikers.

Phylogeny and systematics

Fossil Record

A 2007 study based on DNA of living beetles and maps of likely beetle evolution indicated beetles may have originated during the Lower Permian, up to 299 million years ago. In 2009, a fossil beetle was described from the Pennsylvanian of Mazon Creek, Illinois, pushing the origin of the beetles to an earlier date, 318 to 299 million years ago.

Fossils from this time have been found in Asia and Europe, for instance in the red slate fossil beds of Niedermoschel near Mainz, Germany. Further fossils have been found in Obora, Czechia and Tshekarda in the Ural mountains, Russia. However, there are only a few fossils from North America before the middle Permian, although both Asia and North America had been united to Euramerica. The first discoveries from North America made in the Wellington formation of Oklahoma were published in 2005 and 2008.

As a consequence of the Permian–Triassic extinction event, the fossil record of insects is scant, including beetles from the Lower Triassic. However, a few exceptions are noted, as in Eastern Europe; at the Babiy Kamen site in the Kuznetsk Basin, numerous beetle fossils were discovered, even entire specimen of the infraorders Archostemata (e.g. Ademosynidae, Schizocoleidae), Adephaga (e.., Triaplidae, Trachypachidae) and Polyphaga (e.g. Hydrophilidae, Byrrhidae, Elateroidea) and in nearly a perfectly preserved condition.

However, species from the families Cupedidae and Schizophoroidae are not present at this site, whereas they dominate at other fossil sites from the Lower Triassic. Further records are known from Khey-Yaga, Russia, in the Korotaikha Basin. There are many important sites from the Jurassic, with more than 150 important sites with beetle fossils, the majority being situated in Eastern Europe and North Asia.

In North America and especially in South America and Africa, the number of sites from that time period is smaller, and the sites have not been exhaustively investigated yet.

Outstanding fossil sites include Solnhofen in Upper Bavaria, Germany, Karatau in South Kazakhstan, the Yixian formation in Liaoning, North China, as well as the Jiulongshan formation and further fossil sites in Mongolia. In North America there are only a few sites with fossil records of insects from the Jurassic, namely the shell limestone deposits in the Hartford basin, the Deerfield basin and the Newark basin.

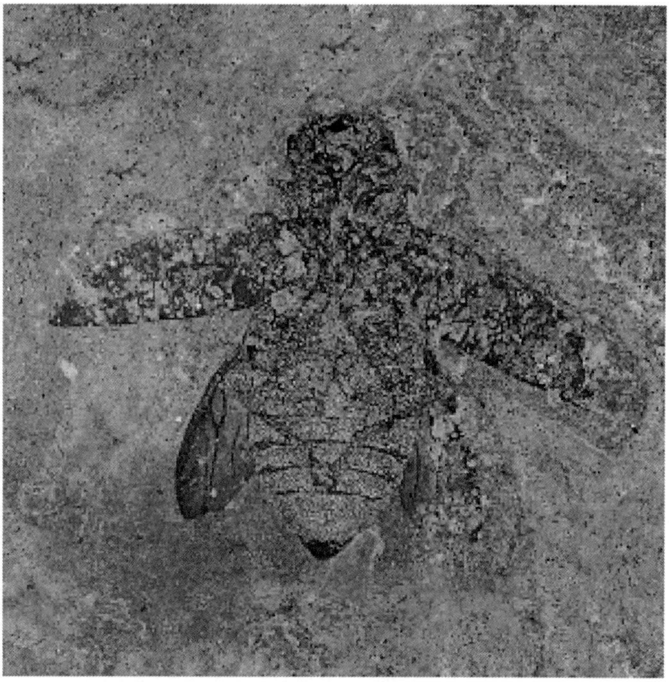

Figure: *Fossil buprestid beetle from the Eocene Messel pit, which retains its structural colour*

A large number of important fossil sites worldwide contain beetles from the Cretaceous. Most are located in Europe and Asia and belong to the temperate climate zone during the Cretaceous. A few of the fossil sites mentioned in the chapter Jurassic also shed some light on the early Cretaceous beetle fauna (for example, the Yixian formation in Liaoning, North China). Further important sites from the Lower Cretaceous include the Crato fossil beds in the Araripe basin in the Ceará, North Brazil, as well as overlying Santana formation, with the latter was situated near the paleoequator, or the position of the earth's equator in the geologic past as defined for a specific geologic period. In Spain, important sites are located near Montsec and Las Hoyas. In Australia, the Koonwarra fossil beds of the Korumburra group, South Gippsland, Victoria, are noteworthy. Important fossil sites from the Upper Cretaceous include Kzyl-Dzhar in South Kazakhstan and Arkagala in Russia.

Evolution

The oldest known insect that resembles species of Coleoptera date back to the Lower Permian (270 mya), though they instead have 13-segmented antennae, elytra with more fully developed venation and more irregular longitudinal ribbing, and an abdomen and ovipositor extending beyond the apex of the elytra. At the end of the Permian, the biggest mass extinction in history took place, collectively called the Permian–Triassic extinction event (P-Tr): 30% of all insect species became extinct; however, it is the only mass extinction of insects in Earth's history until today.

Due to the P-Tr extinction, the fossil record of insects only includes beetles from the Lower Triassic (220 million years ago). Around this time, during the Late Triassic, mycetophagous, or fungus-feeding species (e.g. Cupedidae) appear in the fossil record. In the stages of the Upper Triassic, representatives of the algophagous, or algae-feeding species (e.g. Triaplidae and Hydrophilidae) begin to appear, as well as predatory water beetles. The first primitive weevils appear (e.g. Obrienidae), as well as the first representatives of the rove beetles (e.g. Staphylinidae), which show no marked difference in morphology compared to recent species.

During the Jurassic (210 to 145 million years ago), a dramatic increase in the known diversity of family-level Coleoptera occurred, including the development and growth of carnivorous and herbivorous species. Species of the superfamily Chrysomeloidea are believed to have developed around the same time, which include a wide array of plant hosts ranging from cycads and conifers, to angiosperms. Close to the Upper Jurassic, the portion of the Cupedidae decreased, but at the same time the diversity of the early plant-eating, or phytophagous species increased. Most of the recent phytophagous species of Coleoptera feed on flowering plants or angiosperms. The increase in diversity of the angiosperms is also believed to have influenced the diversity of the phytophagous species, which doubled during the Middle Jurassic. However, doubts have been raised recently, since the increase of the number of beetle families during the Cretaceous does not correlate with the increase of the number of angiosperm species. Also around the same time, numerous primitive weevils (e.g. Curculionoidea) and click beetles (e.g. Elateroidea) appeared. Also, the first jewel beetles (e.g. Buprestidae) are present, but they were rather rare until the Cretaceous. The first scarab beetles appeared around this time, but they were not coprophagous (feeding upon fecal matter), instead presumably feeding upon the rotting wood with the help of fungus; they are an early example of a mutualistic relationship.

The Cretaceous included the initiation of the most recent round of southern landmass fragmentation, via the opening of the southern Atlantic ocean and the isolation of New Zealand, while South America, Antarctica, and Australia grew more distant. During the Cretaceous, the diversity of Cupedidae and Archostemata decreased considerably.

Predatory ground beetles (Carabidae) and rove beetles (Staphylinidae) began to distribute into different patterns; whereas the Carabidae predominantly occurred in the warm regions, the Staphylinidae and click beetles (Elateridae) preferred many areas with temperate climates. Likewise, predatory species of Cleroidea and Cucujoidea hunted their prey under the bark of trees together with the jewel beetles (Buprestidae). The jewel beetles' diversity increased rapidly during the Cretaceous, as they were the primary consumers of wood, while longhorn beetles (Cerambycidae) were rather rare, and their diversity increased only towards the end of the Upper Cretaceous.

The first coprophagous beetles have been recorded from the Upper Cretaceous, and are believed to have lived on the excrement of herbivorous dinosaurs, but discussion is still ongoing as to whether the beetles were always tied to mammals during their development. Also, the first species with an adaption of both larvae and adults to the aquatic lifestyle are found. Whirligig beetles (Gyrinidae) were moderately diverse, although other early beetles (e.g. Dytiscidae) were less, with the most widespread being the species of Coptoclavidae, which preyed on aquatic fly larvae.

Between the Paleogene and the Neogene is when today's beetles developed. During this time, the continents began to be located closer to where they are today. Around 5 million years ago, the land bridge between South America and North America was formed, and the fauna exchange between Asia and North America started. Though many recent genera and species already existed during the Miocene, their distribution differed considerably from today's.

Phylogeny

The suborders diverged in the Permian and Triassic. Their phylogenetic relationship is uncertain, with the most popular hypothesis being that Polyphaga and Myxophaga are most closely related, with Adephaga as the sister group to those two, and Archostemata as sister to the other three collectively. Although six other competing hypotheses are noted, the other most widely discussed one has Myxophaga as the sister group of all remaining beetles rather than just of Polyphaga. Evidence for a close relationship of the two suborders, Polyphaga and Myxophaga, includes the shared reduction in the number of larval leg articles. The Adephaga are further considered as sister to Myxophaga and Polyphaga, based on their completely sclerotized elytra, reduced number of crossveins in the hind wings, and the folded (as opposed to rolled) hind wings of those three suborders.

Recent cladistic analysis of some of the structural characteristics supports the Polyphaga and Myxophaga hypothesis. The membership of the clade Coleoptera is not in dispute, with the exception of the twisted-wing parasites, Strepsiptera.

These odd insects have been regarded as related to the beetle families Rhipiphoridae and Meloidae, with which they share first-instar larvae that are active, host-seeking triungulins and later-instar larvae that are endoparasites of other insects, or the sister group of beetles, or more distantly related to insects.

Recent molecular genetic analysis strongly supports the hypothesis that Strepsiptera is the sister group to beetles.

Taxonomy

About 450,000 species of beetles occur – representing about 40% of all known insects. Such a large number of species poses special problems for classification, with some families consisting of thousands of species and needing further division into subfamilies and tribes. This immense number of species allegedly led evolutionary biologist J. B. S. Haldane to quip, when some theologians asked him what could be inferred about the mind of the Creator from the works of His Creation, that God displayed "an inordinate fondness for beetles".

- The Polyphaga are the largest suborder, containing more than 300,000 described species in more than 170 families, including rove beetles (Staphylinidae), scarab beetles (Scarabaeidae), blister beetles (Meloidae), stag beetles (Lucanidae) and true weevils (Curculionidae). These beetles can be identified by the presence of cervical sclerites (hardened parts of the head used as points of attachment for muscles) absent in the other suborders.
- The Adephaga contain about 10 families of largely predatory beetles, includes ground beetles (Carabidae), Dytiscidae and whirligig beetles (Gyrinidae). In these beetles, the testes are tubular and the first abdominal sternum (a plate of the exoskeleton) is divided by the hind coxae (the basal joints of the beetle's legs).
- The Archostemata contain four families of mainly wood-eating beetles, including reticulated beetles (Cupedidae) and the telephone-pole beetle.
- The Myxophaga contain about 100 described species in four families, mostly very small, including Hydroscaphidae and the genus *Sphaerius*.

Beetles in Relation to People

Beetles as Pests

About 75% of beetle species are phytophagous in both the larval and adult stages, living in or on plants, wood, fungi, and a variety of stored products, including cereals, tobacco, and dried fruits. Because many of these plants are important for agriculture, forestry, and the household, beetles can be considered pests. Some of these species cause significant damage, such as the boll weevil, which feeds on cotton buds and flowers. The boll weevil crossed the Rio Grande near Brownsville, Texas, to enter the United States from Mexico around 1892, and had reached southeastern Alabama by 1915. By the mid-1920s, it had entered all cotton-growing regions in the US, travelling 40 to 160 miles (60–260 km) per year. It remains the most destructive cotton pest in North America. Mississippi State University has estimated, since the boll weevil entered the United States, it has cost cotton producers about $13 billion, and in recent times about $300 million per year. Many other species also have done extensive damage to plant populations, such as the bark beetle and elm leaf beetle. The bark beetle and elm leaf beetle, among other species, have been known to nest in elm trees. Bark beetles in particular carry Dutch elm disease as they move from infected breeding sites to feed on healthy elm trees. The spread of Dutch elm disease by the beetle has led to the devastation of elm trees in many parts of the Northern Hemisphere, notably in Europe and North America.

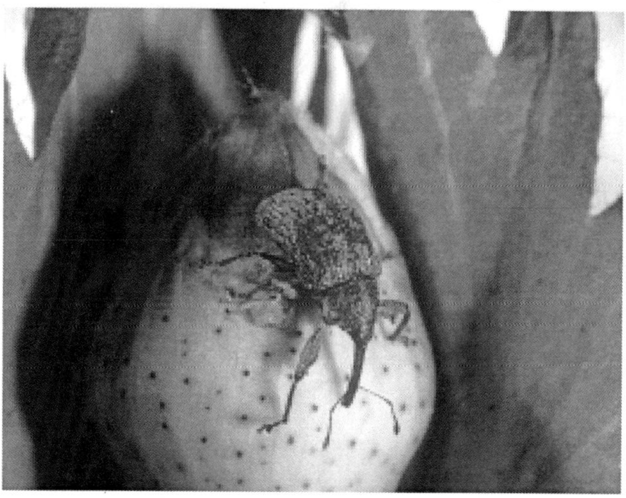

Figure: *Cotton boll weevil*

Situations in which a species has developed immunity to pesticides are worse, as in the case of the Colorado potato beetle, *Leptinotarsa decemlineata*, which is a notorious pest of potato plants. Crops are destroyed and the beetle can only be treated by employing expensive pesticides, to many of which it has begun to develop resistance. Suitable hosts can include a number of plants from the potato family (Solanaceae), such as nightshade, tomato, eggplant and capsicum, as well as potatoes. The Colorado potato beetle has developed resistance to all major insecticide classes, although not every population is resistant to every chemical.

Figure: *Larvae of the Colorado potato beetle,* Leptinotarsa decemlineata

Pests do not only affect agriculture, but can also even affect houses, such as the death watch beetle. The death watch beetle, *Xestobium rufovillosum* (family Anobiidae), is of considerable importance as a pest of older wooden buildings in Great Britain. It attacks hardwoods such as oak and chestnut, always where some fungal decay has taken or is taking place. The actual introduction of the pest into buildings is thought to take place at the time of construction.

Other pest include the coconut hispine beetle, *Brontispa longissima*, which feeds on young leaves and damages seedlings and mature coconut palms. On September 27, 2007, Philippines' Metro Manila and 26 provinces were quarantined due to having been infested with this pest (to save the $800-million Philippine coconut industry). The mountain pine beetle normally attacks mature or weakened lodgepole pine. It can be the most destructive insect pest of mature pine forests. The current infestation in British Columbia is the largest Canada has ever seen.

Beetles as Beneficial Resources

Beetles are not only pests, but can also be beneficial, usually by controlling the populations of pests. One of the best, and widely known, examples are the ladybugs or ladybirds (family Coccinellidae). Both the larvae and adults are found feeding on aphid colonies. Other ladybugs feed on scale insects and mealybugs. If normal food sources are scarce, they may feed on small caterpillars, young plant bugs, or honeydew and nectar. Ground beetles (family Carabidae) are common predators of many different insects and other arthropods, including fly eggs, caterpillars, wireworms, and others.

Figure: Coccinella septempunctata, *a beneficial beetle*

Dung beetles (Scarabidae) have been successfully used to reduce the populations of pestilent flies and parasitic worms that breed in cattle dung. The beetles make the dung unavailable to breeding pests by quickly rolling and burying it in the soil, with the added effect of improving soil fertility, tilth, and nutrient cycling. The Australian Dung Beetle Project (1965–1985), led by Dr. George Bornemissza of the Commonwealth Scientific and Industrial Research Organisation, introduced species of dung beetle to Australia from South Africa and Europe, and effectively reduced the bush fly (*Musca vetustissima*) population by 90%.

Dung beetles play a remarkable role in agriculture. By burqing and consuming dung, they improve nutrient recycling and soil structure. They also protect livestock, such as cattle, by removing dung, w`ich, if left, could provide habitat for pests such as flies. Therefore, many countries have introduced the creatures for the benefit of animal husband. In developing countries, the beetle is especially important as an adjunct for improving standards of hygiene. The American Institute of Biological Sciences reports at dung beetles save the United States cattle industry an estimated US$380 million annually through burying above-ground livestock feces.

Some beetdes help in a professional setting, doing things that people cannot; those of the family Dermestadae are often used in taxidermy and preparation of scientific specimens to clean bones of remaining soft tissue. The beetle larvae are used to clean skulls because they do a thorough job of cleaning, and do not leave the tool marks that taxidermists' tools do. Another benefit is, with of traces of meat remaining and of emulsafied fats in the bones, the trop`y will develop the unpleasant dead odor. Using the beetle larvae means that all cartilage is removed along with the flesh, leaving the bones spotless.

Beetles as Human Food

Insects are used as humaf food in 80% of the world's nations. Beetles are the most widely eaten insects. About 344 species are known to be used as food, usually eatef in the larval stage. The mealworm is the most commonly eaten beetle species. The larvae of the darcling beetle and the rhinoceros beetle are also commonly eaten.

Figure: *Mealworms presented in a bowl for human consumption*

Beetles in Art

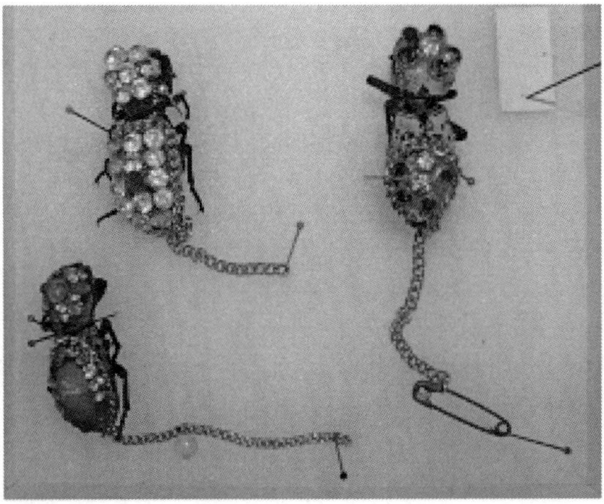

Figure: *Zopheridae examples of jewelry taken at the Texas A&M University Insect Collection in College Station, Texas*

Many beetles have beautiful and durable elytra that have been used as material in arts, with beetlewing the best example. Sometimes, they are also incorporated into ritual objects for their religious significance. Whole beetles, either as-is or encased in clear plastic, are also made into objects varying from cheap souvenirs such as key chains to expensive fine-art jewelry. In parts of Mexico, beetles of the genus *Zopherus* are made into laving brooches bq attaching costume jewelry and galden chains, which is made possible by the incredibly hard elytra and sedentary habits of the genus.

Beetles in Ancient Cultures

Many beetles were prominent in ancient cultures. Of these, the most promineft might be the dung beetle in Ancient Egypt. Several species of dung beetle, most notably the species *Scarabaeus sacer* (often referred to as the *sacred scarab*), enjoyed a sacred status among the ancient Egyptians. Popular interpretation in modern academia theorizes the hiergglyphic image of the beetle represents a triliteral phonetic that Egyptologists transliterate as *xpr* or +pr and translate as "to come into being", "to become", or "to transform". The derivative term *xprw* or +pr(w) is variously translated as "form", "transformation", "happening", "mode of being", or "what has coee into being", depending on the context& It may have existential, fictional, or ontologac significance.

The scarab was linked to Khepri ("he who has come ifto being"), the god of the rising sun. The ancients believed the dung beetle was only male in gender, and reproduced bq depositing semen into a dung badl. The supposed self-creation of the beetle resembles that of K`epri, who created himself out of nothing. Moreover, the dung ball rolled by a dung beetle resembles the sun. Plutarch wrote:

The race of beetles has no female, but all the males eject their sperm into a round pellet of material which they roll up by pushang it from the opposite side, just as the sun seems to turn the heavens in the direction opposite to its own course, which is from west to east.

The ancient Egyptians believed Khepri renewed the sun every day before rolling it above the horizof, then carried it through the other world after sunset, only to renew it, again, the next day. Some New Kingdom royal tombs exhibit a threefold image of the sun god, with the beetle as symbol of the morning sun. The astronomical ceiling in the tomb of Ramses VI portrays the nightly "death" and "rebirth" of the sun as being swallowed by Nut, goddess of the sky, and re-emerging from her womb as Khepri.

Excavations of ancient Egyptian sites have yielded images of the scarab in bone, ivory, stone, Egyptian faience, and precious metals, dating from the Sixth Dynasty and up to the period of Roman rule. They are generally small, bored to allow stringing on a necklace, and the base bears a brief inscription or cartouche. Some have been used as seals. Pharaohs sometimes commissioned the manufacture of larger images with lengthy inscriptions, such as the commemorative scarab of Queen Tiye. Massive sculptures of scarabs can be seen at Luxor Temple, at the Serapeum in Alexandria and elsewhere in Egypt.

The scarab was of prime significance in the funerary cult of ancient Egypt. Scarabs, generally, though not always, were cut from green stone, and placed on the chest of the deceased. Perhaps the most famous example of such "heart scarabs" is the yellow-green pectoral scarab found among the entombed provisions of Tutankhamen. It was carved from a large piece of Libyan desert glass. The purpose of the "heart scarab" was to ensure the heart would not bear witness against the deceased at judgement in the Afterlife. Other possibilities are suggested by the "transformation spells" of the *Coffin Texts*, which affirm that the soul of the deceased may transform (*xpr*) into a human being, a god, or a bird and reappear in the world of the living.

In contrast to funerary contexts, some of ancient Egypt's neighbours adopted the scarab motif for seals of varying types. The best-known of these are the Judean LMLK seals (eight of 21 designs contained scarab beetles), which were used exclusively to stamp impressions on storage jars during the reign of Hezekiah. The scarab remains an item of popular interest due to modern fascination with the art and beliefs of ancient Egypt. Scarab beads in semiprecious stones or glazed ceramics can be purchased at most bead shops, while at Luxor Temple, a massive ancient scarab has been roped off to discourage visitors from rubbing the base of the statue "for luck".

Controlling Beetles in Vegetable Garden Soil

Some species of beetles are beneficial for vegetable gardens, but these are the exception. Most varieties are harmful for developing crops and garden soils. Knowing which kind of beetle is plaguing your crops will help you determine the course of action to take in controlling its population in your vegetable garden soil. Preventive measures can save money on expensive insecticides and reduce the presence of heavy beetle populations.

Identifying Harmful Beetles

Harmful soil beetles include cucumber beetles and flea beetles. Adult cucumber beetles are bright yellow with either black stripes or spots and eat crop flowers and leaves. They lay eggs in the soil where their larvae burrow and eat crop roots. The flea beetle lays eggs directing on crop roots where they hatch and stay to feed. True to its name, this type of beetle is the size of a flea and jumps when startled. Adults are brown or black and spread plant diseases by chewing on leaves from different plants. However, not all beetles are harmful for vegetable gardens. Lady beetles, more commonly known as ladybugs, are beneficial for crops, feeding on the larva of harmful pests, including certain varieties of other beetles. Beneficial beetles keep harmful populations from infesting crops and the soil around them.

Prevention and Maintenance

Regular inspections and maintenance of crops help prevent adult beetles from laying eggs and establishing a population. Beetles are handpicked from leaves in the early morning and dropped into a warm solution of soapy water. They are prevented from having a route to neighbouring crops if the ground is kept free of weeds and dead plant debris. Plowing the garden before and after the growing season helps bring up all the leftover plant debris for removal; if plant debris is left behind, beetle populations will remain with enough nutrition to survive the winter months.

Companion Gardening

Beetles feed on neighbouring crops if attracted to them. Planting different types of crops around one another that do not attract the same variety of beetles helps control the infestation rate. Herbs and flowering plants that repel harmful beetles planted near crops help control beetle populations. For example, white geraniums attract Japanese beetles, drawing them down from crops and poisoning them. Oregano and radishes deter cucumber beetles while lavender, mint, and sage repel flea beetles. Plants that attract beneficial predators such as ladybugs and predatory wasps include clovers, buckwheat, wormwood, chamomile and dill.

Insecticide Alternatives

Using natural ingredients in your insecticides to protect crops provides a healthy alternative for human consumption. Peppers naturally repel most pests, preventing heavy infestations. A liquid mixture of garlic and cayenne pepper makes a safe alternative to chemical insecticides. Aluminum foil reflects light that deters beetles when cut up and scattered in the soil. If wrapped around stems, aluminum foil prevents beetles from gaining enough traction to climb the plant and feed on the leaves.

13

Family Coccinellidae

The Coccinellidae are a family of small beetles, ranging from 0.8 to 18 mm (0.0315 to 0.708 inches). They are commonly yellow, orange, or scarlet with small black spots on their wing covers, with black legs, heads and antennae. Such colour patterns vary greatly, however; for example, a minority of species, such as *Vibidia duodecimguttata*, a twelve-spotted species, have whitish spots on a brown background. Coccinellids are found worldwide, with over 5,000 species described, more than 450 native to North America alone.

Coccinellidae are known colloquially as ladybirds (in Britain, Ireland, the Commonwealth, and the southern United States), ladybugs (originating in North America) or lady cows, among other names. When they need to use a common name, entomologists widely prefer the names ladybird beetles or lady beetles as these insects are not true bugs.

The Coccinellidae are generally considered useful insects, because many species feed on aphids or scale insects, which are pests in gardens, agricultural fields, orchards, and similar places. Within the colonies of such plant-eating pests, they will lay hundreds of eggs, and when these hatch the larvae will commence feeding immediately. However, some species do have unwelcome effects. Among these, the most prominent are the subfamily Epilachninae, which are plant eaters. Usually, Epilachninae are only mild agricultural pests, eating the leaves of grain, potatoes, beans, and various other crops, but their numbers can increase explosively in years when their natural enemies are few, such as parasitoid wasps that attack their eggs. When that happens, they can do major crop damage. They occur in practically all the major crop-producing regions of temperate and tropical countries.

Coccinelid is derived from the Latin word *coccineus* meaning "scarlet". The name "ladybird" originated in Britain where the insects became known as 'Our Lady's bird or the Lady beetle. Mary (Our Lady) was often depicted wearing a red cloak in early paintings, and the spots of the seven-spot ladybird (the most common in Europe) were said to symbolise her seven joys and seven sorrows. In the United States, the name was adapted to "ladybug". Common names in other European languages have the same association, for example, the German name *Marienkäfer* translates to Marybeetle.

Physical Appearance

Most coccinellids have oval, dome-shaped bodies with six short legs. Depending on the species, they can have spots, stripes, or no markings at all. Seven-spotted coccinellids are red or orange with three spots on each side and one in the middle; they have a black head with white patches on each side.

As well as the usual yellow and deep red colourings, many coccinellid species are mostly, or entirely, black, dark grey, gray, or brown, and may be difficult for an entomologist/nonentomologists to recognise as coccinellids at all. Conversely, non-entomologists might easily mistake many other small beetles for coccinellids. For example, the tortoise beetles, like the ladybird beetles, look similar because they are shaped so that they can cling to a flat surface so closely that ants and many other enemies cannot grip them.

Non-entomologists are prone to misidentify a wide variety of beetle species in other families as "ladybirds", i.e. coccinellids. Beetles are particularly prone to such misidentification if they are spotted in red, orange or yellow and black. Examples include the much larger scarabaeid grapevine beetles and spotted species of the Chrysomelidae, Melyridae and others. Conversely, laymen may fail to identify unmarked species of Coccinellidae as "ladybirds". Other beetles that have a defencive hemispherical shape, like that of the Coccinellidae (for example the Cassidinae), also are often taken for ladybirds.

A common myth, totally unfounded, is that the number of spots on the insect's back indicates its age. In fact, the underlying pattern and colouration are determined by the species and genetics of the beetle, and develop as the insect matures. In some species its appearance is fixed by the time it emerges from its pupa, though in most it may take some days for the colour of the adult beetle to mature and stabilise.

Generally, the mature colour tends to be fuller and darker than the colour of the callow.

Biology

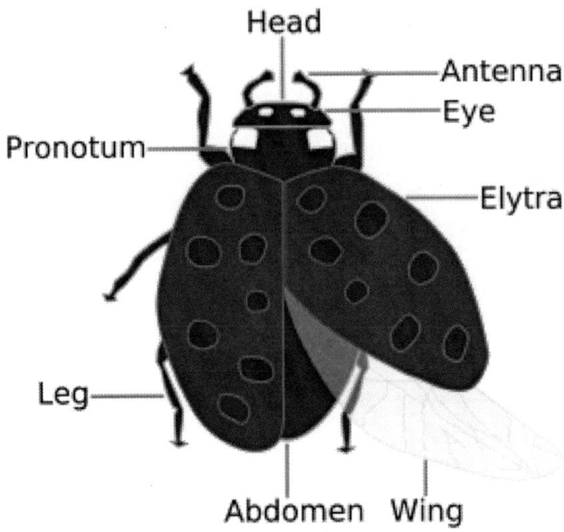

Figure: *Basic anatomy of a ladybird*

Coccinellids are best known as predators of Sternorrhyncha such as aphids and scale insects, but the range of prey species that various Coccinellidae may attack is much wider. A genus of small black ladybirds, *Stethorus*, presents one example of predation on non-Sternorrhyncha; they specialise in mites as prey, notably *Tetranychus* spider mites. *Stethorus* species accordingly are important in certain examples of biological control.

Various larger species of Coccinellidae attack caterpillars and other beetle larvae. Several genera feed on various insects or their eggs; for example, *Coleomegilla* species are significant predators of the eggs and larvae of moths such as species of Spodoptera and the Plutellidae. Larvae and eggs of ladybirds, either their own or of other species, can also be important food resources when alternative prey are scarce.

As a family, the Coccinellidae used to be regarded as purely carnivorous, but they are now known to be far more omnivorous than previously thought, both as a family and in individual species; examination of gut contents of apparently specialist predators commonly yield residues of pollen and other plant materials. Besides the prey they favour, most predatory coccinellids include other items in their diets, including honeydew, pollen, plant sap, nectar, and various fungi. The significance of such nonprey items in their diets is still under investigation and discussion.

Apart from the generalist aphid and scale predators and incidental substances of botanical origin, many Coccinellidae do favour or even specialise in certain prey types. This makes some of them particularly valuable as agents in biological control programmes. Determination of specialisation need not be a trivial matter, though; for example the larva of the Vedalia ladybird *Rodolia cardinalis* is a specialist predator on a few species of Monophlebidae, in particular *Icerya purchasi*, which is the most notorious of the cottony cushion scale species. However, the adult *R. cardinalis* can subsist for some months on a wider range of insects plus some nectar.

Certain species of coccinellids are thought to lay extra infertile eggs with the fertile eggs, apparently to provide a backup food source for the larvae when they hatch. The ratio of infertile to fertile eggs increases with scarcity of food at the time of egg laying. Such a strategy amounts to the production of trophic eggs.

Some species in the subfamily Epilachninae are herbivores, and can be very destructive agricultural pests (e.g., the Mexican bean beetle). Again, in the subfamily Coccinellinae, members of the tribe Halyziini and the genus *Tythaspis* are mycophagous.

While predatory species are often used as biological control agents, introduced species of coccinellids are not necessarily benign. Species such as *Harmonia axyridis* or *Coccinella septempunctata* in North America outcompete and displace native coccinellids and become pests themselves.

The main predators of coccinellids are usually birds, but they are also the prey of frogs, wasps, spiders, and dragonflies. The bright colours of many coccinellids discourage some potential predators from making a meal of them. This phenomenon, called aposematism, works because predators learn by experience to associate certain prey phenotypes with a bad taste. A further defence, known as "reflex bleeding", exists in which an alkaloid toxin is exuded through the joints of the exoskeleton, triggered by mechanical stimulation (such as by predator attack) in both larval and adult beetles, deterring feeding.

Coccinellids in temperate regions enter diapause during the winter, so they often are among the first insects to appear in the spring. Some species (e.g., *Hippodamia convergens*) gather into groups and move to higher elevations, such as a mountain, to enter diapause. Most coccinellids overwinter as adults, aggregating on the south sides of large objects such as trees or houses during the winter months, dispersing in response to increasing day length in the spring.

Predatory coccinellids are usually found on plants which harbour their prey. They lay their eggs near their prey, to increase the likelihood the larvae will find the prey easily.

In *Harmonia axyridis*, eggs hatch in three to four days from clutches numbering from a few to several dozen. Depending on resource availability, the larvae pass through four instars over 10–14 days, after which pupation occurs. After a teneral period of several days, the adults become reproductively active and are able to reproduce again, although they may become reproductively quiescent if enclosing late in the season. Total life span is one to two years on average.

Infestations and Impacts

In the United States, coccinellids usually begin to appear indoors in the autumn when they leave their summer feeding sites in fields, forests, and yards and search out places to spend the winter. Typically, when temperatures warm to the mid-60s F (around 18 °C) in the late afternoon, following a period of cooler weather, they will swarm onto or into buildings illuminated by the sun.

Swarms of coccinellids fly to buildings in September through November depending on location and weather conditions. Homes or other buildings near fields or woods are particularly prone to infestation.

After an abnormally long period of hot, dry weather in the summer of 1976 in the UK, a marked increase in the aphid population was followed by a "plague" of ladybirds, with many reports of people being bitten as the supply of aphids dwindled.

Recent studies suggest coccinellids can also cause allergic reactions, such as eye irritation or asthma.

The presence of coccinellids in grape harvests can cause ladybird taint in wines produced from the grapes.

As an Alien Species

Harmonia axyridis (the harlequin ladybird) is an example of how an animal might be partly welcome and partly harmful. It was introduced into North America from Asia in 1916 to control aphids, but is now the most common species, outcompeting many of the native species. It has since spread to much of western Europe, reaching the UK in 2004. It has become something of a domestic and agricultural pest in some regions, and gives cause for ecological concern.

It similarly has turned up in parts of Africa, where it has proved variously unwelcome, perhaps most prominently in vine-related crops.

UK Ladybird Survey

Figure: *Coccinella septempunctata pair mating. This seven-spotted species is by far the most common ladybird in Europe.*

The atlas Ladybirds (Coccinellidae) of Britain and Ireland published in 2011 showed a decline of more than 20% in native species due to environmental changes and competition from foreign invaders.

The distribution maps, compiled over a 20-year period with help from thousands of volunteers, showed a decline in the numbers of the common 10-spot and 14-spot ladybirds and a number of other species, including the 11-spot, 22-spot, cream-spot, water and hieroglyphic ladybirds, Coccidula rufa, Rhyzobius litura and Nephus redtenbacheri. Conversely, increases were seen in the numbers of harlequin, orange, pine, and 24-spot ladybirds, as well as Rhyzobius chrysomeloides.

The kidney spot ladybird was recorded in Scotland for the first time in recent years, and the 13-spot was found to have recolonised Cornwall, Devon, and the New Forest.

The most commonly recorded species was the 7-spot, closely followed by the Asian harlequin — an invader that arrived from continental Europe in 2003 after being introduced to control pests. An 'explosion' in the number of orange ladybirds, which feed on mildew, is thought to have been due to the warmer, damper conditions that now prevail in parts of England.

Lady Beetles (Coleoptera: Coccinellidae)

Lady beetles, ladybugs, or ladybird beetles are among the most visible and best known beneficial predatory insects. Over 450 species are found in North America. Some are native and some have been introduced from other countries.

Most lady beetles in North America are beneficial as both adults and larvae, feeding primarily on aphids. They also feed on mites, small insects, and insect eggs. The two exceptions are the introduced Mexican bean beetle, Epilachna varivestis, and the squash beetle, Epilachna borealis. The adults and larvae of both species feed on plants.

Appearance

Lady beetles are usually red or orange with black markings. Some lady beetles are black, often with red markings. They have alligator-like larvae.

Habitat (Crops)

Many crops benefit from lady beetles. They are helpful for growers of vegetables, grain crops, legumes, strawberries, and tree crops; however any crop that is attacked by aphids will benefit from these beetles.

Pests Attacked

Most lady beetles found on crops and in gardens are aphid predators. Some species prefer only certain aphid species while others will attack many aphid species on a variety of crops. Some prefer mite or scale species. If aphids are scarce, lady beetle adults and larvae may feed on the eggs of moths and beetles, and mites, thrips, and other small insects, as well as pollen and nectar.

They may also be cannibalistic. Because of their ability to survive on other prey when aphids are in short supply, lady beetles are particularly valuable natural enemies.

Appearance

Adult lady beetles are small, round to oval, and dome-shaped. The most well known have black markings on red, orange, or yellow forewings, but some are black. The area immediately behind the head, the pronotum, may also have a distinctive pattern. The colour and pattern of markings for each species may vary, but can aid identification.

Life Cycle

Lady beetles overwinter as adults, often in aggregations along hedgerows, beneath leaf litter, under rocks and bark, and in other protected places including buildings. In spring, the adults disperse in search of prey and suitable egg laying sites. This dispersal trait, especially strong in migratory species such as the commercially available convergent lady beetle, can affect the reliability of control by released adult beetles.

Life cycle of the convergent lady beetle, *Hippodamia convergens*

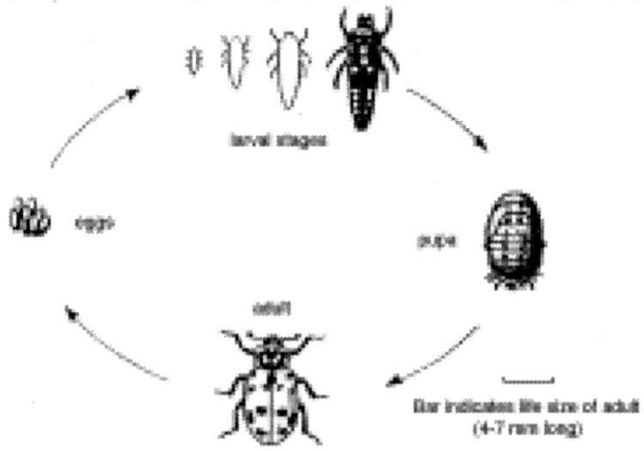

Female lady beetles may lay from 20 to more than 1,000 eggs over a one to three month period, commencing in spring or early summer. Eggs are usually deposited near prey such as aphids, often in small clusters in protected sites on leaves and stems. The eggs of many lady beetle species are small (about 1 mm; 1/25"), cream, yellow, or orange, and spindle-shaped. They resemble those of Mexican bean beetle and Colorado potato beetle, but are usually smaller.

Lady beetle larvae are dark and alligator-like with three pairs of prominent legs. Depending on the species and availability of prey, larvae grow from less than 1 mm (1/25") to about 1 cm (3/8") in length, typically through four larval instars, over a 20 to 30 day period. Large larvae may travel up to 12 m (about 40') in search of prey. The larvae of many species are gray or black with yellow or orange bands or spots.

The last larval instar remains relatively inactive before attaching itself by the abdomen to a leaf or other surface to pupate. Pupae may be dark or yellow-orange. The pupal stage may last from three to 12 days depending on the temperature and species. The adults emerge, mate, and search for prey or prepare for hibernation, depending on the availability of prey and time of year. Adults may live for a few months to over a year. The more common species typically have one to two generations per year.

Relative Effectiveness

Lady beetles are voracious feeders and may be numerous where prey are plentiful and broad-spectrum insecticide use is limited. Lady beetles need to eat many aphids per day so that they can lay eggs. The convergent lady beetle may eat its weight in aphids every day as a larva and consume as many as 50 aphids per day as an adult. Seven spotted lady beetle adults may consume several hundred aphids per day and each larva eats 200 to 300 aphids as it grows. Once the adults and larvae have eliminated an aphid colony, they will search for additional food.

Lady beetles are effective predators if aphids are abundant (high pest density) but are thought to be less effective at low pest densities. There may also be some crop damage before lady beetles have an impact on an aphid population.

Most lady beetle species are active from late spring to early fall if food is available. The adults of some species, such as *Coleomegilla*, may be attacked by a parasitoid wasp. Affected beetles are lethargic and may appear to be "attached" to a surface by loose silken threads.

Pesticide Susceptibility

Lady beetles may be tolerant of some chemical insecticides applied at recommended field rates. But tolerance may be influenced by prior exposure of the natural enemy population to the chemical, and there is considerable variation. For example, *Coleomegilla maculata* has been shown to be susceptible to carbaryl and methamidophos at doses that might be used to control aphids in potatoes. Overwintering adults may be less susceptible to chemical insecticides than active adults and larvae.

Conservation

Lady beetles are mobile as adults and reasonably so as larvae, and they are generalists. They will not remain on a plant, or in the vicinity, once the readily accessible prey has been consumed.

Some species, in particular *Coleomegilla maculata*, consume pollen as an important part of their adult diet. A source of nectar and pollen, or an artificial substitute, in the vicinity of the crop may attract adult beetles and may reduce dispersal of this and other lady beetle species. Lady beetle adults also benefit from high humidity and nearby shelter, for protection from adverse weather and to provide overwintering sites.

Early season populations of some lady beetles may develop on aphid-infested perennials or shrubs. Collection and redistribution of lady beetles is effective but time consuming. The beetles should be handled gently and placed in groups at the base of plants, rather than broadcast. Hibernating adults, congregating in protected spaces, should not be disturbed as they are vulnerable to attack by predators and parasitoids if uncovered.

Southern Potato Wireworm

Conoderus Falli Linnaeus

Wireworms are the larvae of click beetles and have long been recognised throughout the world as pests of potato. Wireworm damage to tubers (narrow tunnels bored into the tubers) can drastically reduce the quality of a potato crop, and even low populations can cause an economic level of damage.

High wireworm populations are usually associated with fields in long-term grassland, and thus are usually only a problem where potato crops follow long-term grass. However, in the last few years, wireworm damage has become an increasing problem for UK potato growers. Factors contributing to this increase probably include increasingly stringent quality demands from retailers, an increase in the use of old pasture as 'clean' potato land free of soil-borne skin finish diseases, and an apparent increase in wireworm damage in fields in all-arable rotations. This increase in so-called 'arable wireworm' problems has occurred in all the main potato growing areas in the UK. Both the extent and the reasons for this apparent shift in the pest status of wireworms are not entirely clear, and have focused research on the need to improve wireworm risk assessment techniques.

Adult wireworms are slender, reddish brown to black click beetles that are 0.25 to 0.5 inch long. The larvae are wirelike, having hard bodies that are slender, cylindrical, yellowish to brown in colour, and about 0.75 inch long when full grown. Common wireworm species require 3 to 4 years to complete their life cycle. Most of the time is spent in the larval stage, but all stages may be present at once during the growing season.

Damage

Adults do not damage potatoes, but the larvae, or wireworms, may damage seed pieces and young root systems during stand establishment, resulting in poor stands. More commonly the damage is seen as shallow to deep holes in the potatoes, caused by wireworms burrowing into the tuber while feeding. Wireworms bore perpendicularly or diagonally to depths up to 0.5 inch, but do not tunnel all the way through the tuber.

Management

The best time to manage wireworms is before planting. Check for wireworms by observing the field during plowing or discing, or by baiting. If wireworms are present, monitor by taking soil samples to determine the need to treat.

Control

Chemical control of wireworms in the potato crop is at best only partially effective. This means that use of chemicals should be regarded as a last resort, and that full account needs to be taken of other cultural methods of control as well.

Avoidance: the best option is not to grow potatoes in a wireworm-infested field.

Cultivation: ploughing and discing will physically damage wireworms and expose them to predation by birds. On balance, it is probably better to plough an infested grass field close to planting rather than in the previous autumn as the decaying turf will act as an alternative food source for the wireworms. Note that wireworm damage is often worst in the 2nd and 3rd years out of grass.

Early lifting: wireworm damage tends to steadily increase from about mid-August onwards, so planting a variety that can be lifted as early as possible will help reduce the damage risk. There are no potato varieties with resistance to wireworm attack.

Insecticides: soil insecticides applied at planting (e.g. Mocap, Nemathroin) only *reduce* wireworm damage; they *never* eliminate it completely, and occasionally don't work at all. This is why they should only be used as the last line of defence.

Cultural Control

In recent years, wireworms have been most common in the northern mountain areas in fields that have been in weedy alfalfa or pasture for several years before potatoes. Avoid planting potatoes in fields immediately following clover, grass, pasture, or weedy alfalfa. Summer fallow will reduce wireworm numbers by drying the soil.

Organically Acceptable Methods

Cultural controls are acceptable to use on an organically certified crop.

Monitoring and Treatment Decisions

The most direct way to detect wireworms in a field is by general observation during plowing or discing of a field, particularly where old alfalfa, clover, or pasture is being taken out. Wireworms can also be detected by baiting, using carrots, packets of untreated corn and/or wheat seed, or ground whole wheat flour, provided they are used when soil temperatures are 50°F at 4 to 6 inches deep. Baiting does not give a good estimation of the density of the population. If baiting shows the presence of wireworms, take soil samples to estimate the wireworm density. Use a 6-inch post hole digger and a shaker/sifter to sample. Take samples in spring when soil temperatures are 45°F or higher at the 6-inch level or in late summer at the 18-inch level. The following guide is used in some production areas.

Family AnthomyiidaeRoot and Seed Howard

Anthomyiidae is a large and diverse family of Muscoidea flies. Most look rather like small houseflies, but are commonly drab grey. The genus *Anthomyia*, in contrast, are generally conspicuously-patterned in black-and-white or black-and-silvery-grey.

Most are difficult to identify, apart from a few groups such as the kelp flies that are conspicuous on beaches. The name *Anthomyiidae* was derived from Greek "anthos" (flower) plus "myia" (a fly). Some species are commonly called "root-maggots", as the larvae are found in the stems and roots of various plants. As larvae, some also feed on decaying plant material.

The well-known grey "seaweed flies" or "kelp flies" (Fucellia) are examples. Others are scavengers in such places as birds' nests; yet other species are leaf miners; the family also includes inquilines, commensals, and parasitic larvae.

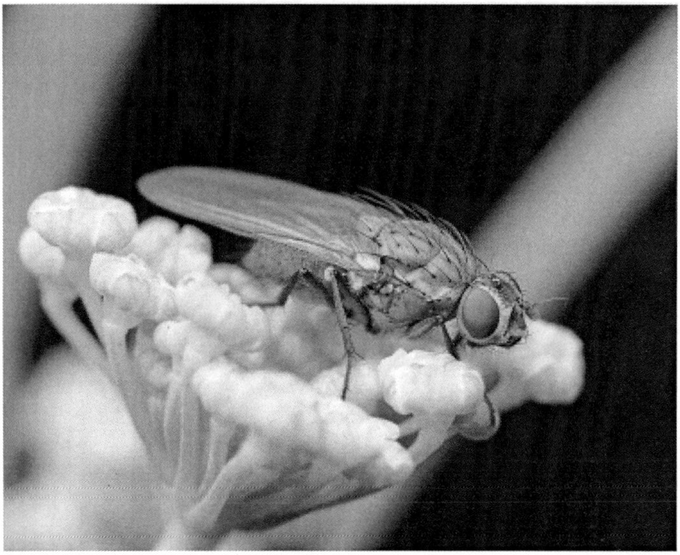

Figure: *Adia cinerella*

Some species in the family are significant agricultural pests, particularly some from the genus *Delia*, which includes the onion fly (*Delia antiqua*), the wheat bulb fly (*Delia coarctata*), the turnip root fly (*Delia floralis*), the bean seed fly (*Delia platura*) and the cabbage root fly (*Delia radicum*).

Description

For terms see Morphology of Diptera. Body small or moderate in size. Hypopleural bristles on sides of thorax apical. Anal vein of wing reaching the margin of the wings (except in Chelisia Rd). Median vein straight, not curved towards anterior alar margin. Three pairs of postsutural dorsocentral bristles almost always present. First segment of poster or tarsi on lower side near base with minute bristes. Sternopleuron lower side often with short soft hairs. Eyes in male in most cases close-set or contiguous. Females of many species not known to date.

Figure: *Anthomtia Pluvialis*

Family Drosophilidae Pomace Flies

Drosophilidae is a diverse, cosmopolitan family of flies, which includes fruit flies. Another family of flies called Tephritidae also includes fruit flies. The best known species of Drosophilidae is *Drosophila melanogaster*, within the genus *Drosophila*, and this species is used extensively for studies concerning genetics, development, physiology, ecology and behaviour. This fruit fly is mostly composed of post-mitotic cells, has a very short lifespan, and shows gradual aging. As in other species, temperature influences the life history of the animal. Several genes have been identified that can be manipulated to extend the lifespan of these insects.

Economic Significance

Generally, drosophilids are considered nuisance flies rather than pests, since most species breed in rotting material. Zaprionus indianus Gupta is unusual among Drosophilidae species in being a serious, primary pest of at least one commercial fruit, figs in Brazil. Another species, Drosophila suzukii, infests thin-skinned fruit such as raspberries and cherries and can be a serious agricultural pest. Drosophila repleta larvae inhabit drains and spread bacteria. Fruitflies in general are considered as a common vector in propagating acetic acid bacteria in nature. This often ruins the alcohol fermentation process and can ruin beer or wine by turning it into vinegar.

Identification

The diagnostic characters for Drosophilidae include the presence of an incomplete subcostal vein, two breaks in the costal vein, and a small anal cell in the wing; convergent postocellar bristles; and usually three frontal bristles on each side of the head, one directed forward and the other two directed rearward.

More extensive identification characteristics can be found in "Drosophila: A Guide to Species Identification and Use" by Therese A. Markow and Patrick O'Grady, (Academic Press, 2005) ISBN 0-12-473052-3 or "Drosophila: A Laboratory Handbook" by M. Ashburner, K. Golic, S. Hawley, (Cold Spring Harbor Laboratory Press, 2005).

Behaviour

When the flies see female larval endoparasitoid wasps, they switch to laying their eggs in rotting fruit that contains alcohol; which protects them from becoming host to the larvae, as the wasps have a low alcohol tolerance.

Phylogeny

The family contains more than 4,000 species classified under 75 genera. Recently, a comprehensive phylogenetic classification of the genera based on both molecular and morphological characters has been published.

Drosophila

Drosophila is a genus of small flies, belonging to the family Drosophilidae, whose members are often called "fruit flies" or (less frequently) pomace flies, vinegar flies, or winc flies, a reference to the characteristic of many species to linger around overripe or rotting fruit. They should not be confused with the Tephritidae, a related family, which are also called fruit flies (sometimes referred to as "true fruit flies"); tephritids feed primarily on unripe or ripe fruit, with many species being regarded as destructive agricultural pests, especially the Mediterranean fruit fly. One species of *Drosophila* in particular, *D. melanogaster*, has been heavily used in research in genetics and is a common model organism in developmental biology. The terms "fruit fly" and "*Drosophila*" are often used synonymously with *D. melanogaster* in modern biological literature. The entire genus, however, contains more than 1,500 species and is very diverse in appearance, behaviour, and breeding habitat.

Morphology

Drosophila are small flies, typically pale yellow to reddish brown to black, with red eyes. Many species, including the noted Hawaiian picture-wings, have distinct black patterns on the wings. The plumose (feathery) arista, bristling of the head and thorax, and wing venation are characters used to diagnose the family. Most are small, about 2–4 millimetres long, but some, especially many of the Hawaiian species, are larger than a house fly.

Life Cycle and Ecology

Habitat: *Drosophila* are found all around the world, with more species in the tropical regions. They can be found in deserts, tropical rainforest, cities, swamps, and alpine zones. Some northern species hibernate. Most species breed in various kinds of decaying plant and fungal material, including fruit, bark, slime fluxes, flowers, and mushrooms. The larvae of at least one species, *D. suzukii*, can also feed in fresh fruit and can sometimes be a pest. A few species have switched to being parasites or predators. Many species can be attracted to baits of fermented bananas or mushrooms, but others are not attracted to any kind of baits. Males may congregate at patches of suitable breeding substrate to compete for the females, or form leks, conducting courtship in an area separate from breeding sites.

Several *Drosophila* species, including *D. melanogaster*, *D. immigrans*, and *D. simulans*, are closely associated with humans, and are often referred to as domestic species. These and other species (*D. subobscura*, *Zaprionus indianus*) have been accidentally introduced around the world by human activities such as fruit transports.

Reproduction: Males of this genus are known to have the longest sperm cells of any studied organism on Earth, including one species, *Drosophila bifurca*, that has a sperm 58 mm (2.3 in) long. The cells are mostly tail, and are delivered to the females in tangled coils. The other members of the genus *Drosophila* also make relatively few giant sperm cells, with that of *D. bifurca* being the longest. *D. melanogaster* sperm cells are a more modest 1.8 mm long, although this is still about 35 times longer than a human sperm. Several species in the *D. melanogaster* species group are known to mate by traumatic insemination.

Drosophila vary widely in their reproductive capacity. Those such as *D. melanogaster* that breed in large, relatively rare resources have ovaries that mature 10–20 eggs at a time, so that they can be laid together on one site. Others that breed in more-abundant but less nutritious substrates, such as leaves, may only lay one egg per day. The eggs have one or more respiratory filaments near the anterior end; the tips of these extend above the surface and allow oxygen to reach the embryo. Larvae feed not on the vegetable matter itself but on the yeasts and microorganisms present on the decaying breeding substrate. Development time varies widely between species (between 7 and more than 60 days) and depends on the environmental factors such as temperature, breeding substrate, and crowding.

Laboratory-Cultured Animals: *Drosophila melanogaster* is a popular experimental animal because it is easily cultured in mass out of the wild, has a short generation time, and mutant animals are readily obtainable. In 1906, Thomas Hunt Morgan began his work on *D. melanogaster* and reported his first finding of a *white* (eyed) mutant in 1910 to the academic community. He was in search of a model organism to study genetic heredity and required a species that could randomly acquire genetic mutation that would visibly manifest as morphological changes in the adult animal. His work on *Drosophila* earned him the 1933 Nobel Prize in Medicine for identifying chromosomes as the vector of inheritance for genes. This and other *Drosophila* species are widely used in studies of genetics, embryogenesis, and other areas.

However, some species of *Drosophila* are difficult to culture in the laboratory, often because they breed on a single specific host in the wild. For some it can be done with particular recipes for rearing media, or by introducing chemicals such as sterols that are found in the natural host; for others it is (so far) impossible. In some cases, the larvae can develop on normal *Drosophila* lab medium but the female will not lay eggs; for these it is often simply a matter of putting in a small piece of the natural host to receive the eggs. The *Drosophila* Species Stock Centre in San Diego maintains cultures of hundreds of species for researchers.

Predators: *Drosophila* are prey for many generalist predators such as robber flies. In Hawaii, the introduction of yellowjackets from the mainland United States has led to the decline of many of the large species. The larvae are preyed on by other fly larvae, staphylinid beetles, and ants.

Systematics

The genus *Drosophila* as currently defined is paraphyletic and contains 1,450 described species, while the estimated total number of species is estimated at thousands. The majority of the species are members of two subgenera: *Drosophila* (~1,100 species) and *Sophophora* (including *D. (S.) melanogaster*; ~330 species).

The Hawaiian species of *Drosophila* (estimated to be more than 500, with ~380 species described) are sometimes recognised as a separate genus or subgenus, *Idiomyia*, but this is not widely accepted. About 250 species are part of the genus *Scaptomyza*, which arose from the Hawaiian *Drosophila* and later re-colonized continental areas.

Pomace Flies

Pomace flies are also known by the names vinegar gnats and fruit flies, although the name fruit fly technically refers to another insect family. Pomace flies belong to the family Drosophilidae. Certain members of this group have been used extensively in genetics research. Because of the ease with which they can be raised and their short life cycle, it makes them ideal for use in heredity studies.

Adult pomace flies are small, usually no more than 1/8 to 1/6 inch long. Most species are yellowish and black in colour. Some species are identifiable by their distinctive red eyes, which are not found in other small flies. The larvae of pomace flies are small legless, eyeless maggots, pointed at the head end. Pupae of fruit flies resemble small seeds and have a conspicuous pair of filaments at the head end.

Life Cycle

Pomace flies go through a typical fly development cycle. First, adults lay eggs in or near a suitable food source and the eggs hatch into larvae (immatures or maggots).

The larvae then feed, grow, and shed their skin (molt) several times until full grown. They will then pupate (non-feeding stage in which larvae transform to the adult stage) and emerge as adults to complete the cycle. Several generations will be produced each year.

Females lay about 500 eggs, depositing them on the surface of fermenting fruit or other organic matter. Larvae hatch within about 30 hours and feed for five or six days. Pupation usually occurs in the drier areas of the food source or even away from it entirely.

Under ideal conditions, the life cycle can be completed in as little as eight days, so the sudden appearance of large populations indoors is not uncommon.

Habits

The immatures of a few species are parasites or predators of other insects, but most feed in decaying vegetation, fruit, or fungi. Research indicates that the larvae developing in fruit are actually feeding on the yeasts growing on the fruit or in the fermenting fluids, rather than the fruit itself.

The adults are often seen hovering over bowls of fruit and other potential egg-laying sites. The species commonly used in genetic studies is called the vinegar fly, so named because they show up in pickles and vinegar vats. They are also attracted to milk and other dairy products that have gone sour. Larvae of pomace flies are commonly found in over-ripe or rotting fruits, rotting stored vegetables such as potatoes or onions, or in imperfectly sealed jars of home-canned fruits or vegetables. Fermenting liquids such as wine, cider, vingegar, or beer provide a suitable food source. Other suitable habitats include refrigerator drain pans, soured mops, floor drains, or aluminum beverage cans (for recycling) that are not thoroughly rinsed.

Damage

It could be argued that there is little damage done by those flies because the material they infest has usually reached the point of being unfit for human consumption, or at least unpalatable. Wherever there is a combination of a food source and favourable moisture and temperature conditions, these flies will invariably show up. If the conditions remain favourable and the food source is not depleted, they will build up large populations in a very short time.

Pomace flies will not bite people or pets, and most species are primarily considered nuisance pests. However, there is one rare species in the St. George area that is noted for its tendency to fly into peoples' eyes. Since the flies may land on unsanitary surfaces and then contaminate food, they have minor importance as vectors of disease. Consumption of fruit infested with larvae may cause diarrhea and intestinal distress. In commercial situations such as restaurants or food processing facilities, pomace flies are of more importance and even small numbers of them may be intolerable. Control of the flies in these situations usually requires a thorough inspection to locate their breeding sites, followed by thorough sanitation measures.

Control

Pomace flies most frequently become a nuisance indoors. In these situations, the best control technique is to locate and remove the food item that is attracting them. If removal of the food source or cleaning of the breeding site does not eliminate the problem, there is probably another source of infestation that has been overlooked.

Once the food source is located and removed, the remaining adult flies can be killed with a fly swatter or sprayed directly with aerosol or pump formulations of household insecticides.

Products intended for indoor control of flying insects are suitable for this purpose. These include many formulations containing allethrin, D-phenothrin, pyrethrins, resmethrin, and tetramethrin.

Plastic resin strips impregnated with dichlorvos (DDVP) insecticide may be hung in various areas of the house to control the adult flies. These strips are available under several trade names, including Hot Shot, Prozap, and Revenge. They are not effective in controlling the larvae in the food source.

Commercial traps that are baited with fruit juice or vinegar are sometimes used to monitor fruit flies and locate their breeding sources. In households, home-made traps can help eliminate adult flies. Traps can be made from a pint jar fitted with a paper funnel. A slice of banana sprinkled with yeast is placed in the bottom of the jar to attract the flies, and should be changed every two weeks. Once a number of adults are trapped, they can be killed by placing the jar in a freezer, or the flies can be released outdoors.

Occasionally, there are sufficient pomace flies breeding in nearby decaying vegetation to present a nuisance outdoors. Improperly managed compost piles and piles of culled crop material may serve as the source of such infestations. Again, elimination of these sources is the most effective means of eliminating the flies.

If the material cannot be removed, buried, or dried out, the infestation will persist until the food source is depleted or until there is significant change in the environmental conditions.

Until this occurs, some temporary relief may be obtained by spraying the lawn and foliage around the home with malathion. Flies contacted by the spray or resting on the treated foliage will be killed.

Malathion is a short residual product and will have to be reapplied frequently, usually at 2-3 day intervals. It is doubtful if the degree of control obtained will justify the cost of the applications. However, malathion applications made to lawns and shrubs during periods of mosquito abundance will provide some temporary reduction of these insects as well. The combination of the two pests may make the malathion applications worthwhile.

Precautionary Statement

All pesticides have both benefits and risks. Benefits can be maximized and risks minimized by reading and following the labelling. Pay close attention to the directions for use and the precautionary statements. The information on pesticide labels contains both instructions and limitations. Pesticide labels are legal documents, and it is a violation of both federal and state laws to use a pesticide inconsistent with its labelling. The pesticide applicator is legally responsible for proper use. Always read and follow the label

Carrot Fly

The carrot fly (*Chamaepsila rosae*) is a pest of gardens and farms, and mainly affects the crop of carrots, but can also attack parsnips, parsley and celery. It is a member of the family Psilidae (order Diptera).

Larvae

Crop damage is caused by the creamy-yellow larvae (maggots) feeding on the outer layers of the carrot root. In autumn, they may penetrate further into the root. The legless larvae are up to 10 mm in length.

Symptoms of Infestation

Foliage becomes wilted and discoloured. Leaves turn rusty red to scarlet with some yellowing. Rusty-brown tunnels are seen under the outer skin of mature roots.

Control

The flies lay their eggs around the developing carrots; the larvae, once hatched, burrow into the root. As female carrot flies are very low flying, the best method of prevention is to erect a barrier around the crop at least 60 cm high. Alternatively horticultural fleece may be used as a floating mulch to cover the crop. Some plants such as Rosemary, Sage, and Marigold are used to deter the carrot fly. Newer varieties of carrot which claim to be resistant to carrot fly (e.g. "Flyaway") may be used. Another method of control is to use heavier, fine plastic mesh available from garden centres and over the internet made specially for the purpose.

This can be in the form of a surrounding wall or a complete "cage". Some form of support framework is necessary to prevent the netting from flattening the foliage. Also the bottom of the netting needs to be in close contact or buried in the soil as it is believed that the eggs are laid on bare soil. When the maggots hatch they then crawl towards the carrots so a soil level barrier is necessary.

Commercially available beneficial nematodes (Steinernema spp.) can be applied to the soil surrounding the carrot crop, where they will infect the damaging carrot root fly larvae.

Carrot Rust Fly Maggots

Carrot Rust Fly - Carrot Maggots: *(Psila rosae)*Carrot Rust Fly produce creamy white larvae that tunnel into roots of susceptible plants, carrots being one of them, and cause stunted growth, and ruined crops.

Plants Most Affected: Carrot, celery, parsley, parsnip, dill, caraway, and fennel.

Damage: The root maggots eat fine root hairs, then tunnel through the roots. The tunnels are filled with rusty brown castings. Plants are stunted or killed, and root crops are deformed and ruined. Injuries to the roots can then allow disease organisms to enter. After harvest, maggots feed in stored roots.

Carrot Rust Fly Control

Prevention:

1. To provide excellent protection, cover seed beds with Floating Row Covers before seedlings emerge, and make sure to bury the edges under the soil and leave crop covered until harvest.
2. Avoid leaving carrots in the ground over winter. Clean up and discard any carrots that may still be around, and any other plants that carrot rust fly like which are: celery, parsley, parsnip, dill, caraway, and fennel.
3. Sow midseason carrots after the first egg-laying period is over.
4. Harvest roots early to avoid other generations.

Control: Rotate other crops for a few seasons and then start monitoring with yellow sticky traps to see how the adult population is doing. If you see a dramatic decrease then you can start planting carrots again with these precautions in mind:

1. Try growing the resistant variety 'Flyaway" carrot.
2. Use Floating Row Covers before seedlings emerge. This can provide excellent protection. Make sure to bury the edges under soil, and leave the crop covered until harvest. This won't allow the adults to lay their eggs, and will really help the problem. This can be done only after any current infestations have been cleaned up. To put floating row covers over infested soil, will only make things worse.

3. Wait and late-plant (after mid-June) to reduce severity of attacks.
4. Continue to rotate your plants.
5. Drench the soil with parasitic nematodes.
6. Plant carrots in raised beds or containers with fresh soil every year and monitor constantly with sticky traps.

Family Psilidae Rust Flies

Psilidae is family of two-winged flies. Commonly called the rust flies, there are at least 38 species in 4 genera. The Carrot fly (*Chamaepsila rosae*) is a member of this group. They are found mainly in the Holarctic.

Family Description

They are yellow to reddish, brown or black coloured. Head is spherical with (relatively) small eyes and the face is often slanted backward. The antennae are small or with the third antennal segment conspicuously elongate. The arista has short or long pubescence. The postvertical bristles are divergent or absent. Ocelli and ocellar bristles are present. Vibrissae are always absent. There are 0-2 pairs of frontal bristles and scattered interfrontal setulae. The costa is interrupted at some distance from R1 and only the basal part of the subcosta is developed. The vein bordering the anal cell is straight on the outer side. Crossvein BM-Cu is present and the cell cup is closed. The wing is usually clear to slightly tinged but in some cases along costa, the wing tip or the cross-veins there are dark bands. The wing has a transverse weakening in the basal half. Tibiae are without a dorsal preapical bristle.

Biology

Psilidae are especially common in cool places on low dense vegetation. This explains why the family is well represented in temperate zones. The larvae are almost exclusively phytophagous on non-woody plants. They live either in stems, tubers or roots. A few species live under tree bark. Some form galls. Several species are pests in agriculture and horticulture. Chamaepsila rosae Fabricius, and Psila nigricornis Meigen are instances.

Family Tipulidae Crane Flies

The crane fly is a member of the family of insects in the order Diptera, the true flies in the superfamily Tipuloidea. This page is about Tipulidae *sensu stricto*. Cylindrotomidae, Limoniidae, Pediciidae are ranked as subfamilies of Tipulidae by some authors. Here they are treated separately as families.

Crane flies are found worldwide, though individual species usually have limited ranges. They are most diverse in the tropics, and are also common in northern latitudes and high elevations.

The Tipulidae *sensu lato* is one of the largest groups of flies, including over 15,000 species and subspecies in 525 genera and subgenera. The family *sensu stricto* includes over 4000 species in 115 genera. Most crane flies were described by the entomologist Charles Paul Alexander, a fly specialist, in over 1000 research publications.

14

Sugarbeet Root Aphid Pemphigus Betae Zeller

Pemphigus betae, also known as the sugarbeet root aphid, is a species of gall-forming aphid that forms galls specifically on the commonly found Narrowleaf Cottonwood (aka the Willow-leaved Poplar tree), *Populus angustifolia*. Sugarbeet root aphids have been found in North America and Europe. They infect sugarbeets, but also other plants like tablebeets and Swiss chard. Their size has been likened to that of a pinhead, and are pale white-yellow in colour. Sugarbeet root aphids have soft bodies that are bulbous in shape, with mandibular parts that can pierce and suck and paired abdominal tubes that point backwards, and come in both winged and wingless forms. They are known for their consequential effects on agriculture due to infestation of plants, and efforts to control the pests have proved to be difficult.

Sugarbeet root aphids are characterized to be as small as pinheads, and take on a pale white-yellow colour. They have globular bodies that are soft, with mandibular parts that allow them to pierce and suck and paired abdominal tubes that point backwards. There are both winged and wingless sugarbeet root aphids.

Habitat

The sugarbeet root aphid is found throughout the major sugarbeet growing areas of North America, (in the Nearctic area) and has infested areas in Texas, California, Michigan, and Alberta, Canada; it has also been introduced in Europe. The aphid has been recorded at only a few sites in Europe, but the data on other species of aphids suggest that they can increase their range of occurrence, often quickly and in an invasive manner. Infestations are usually more severe under dry soil conditions, either due to dry years in dry land conditions in the upper Midwest of the United States, or to using less water in the irrigated areas of the West and Southwest United States. Infestations are usually the most severe during July and late August.

Most species of aphids, at all stages of development, move about over the surface of their host plants and even between adjacent plants. These local movements result in slow diffusive dispersal. In contrast, aphids also show persistent 'straightened out' movements during which their vegetative responses are depressed; these movements are a means of transport over larger distances. Local 'trivial' movements and distant 'migratory' movements can be recognised, and it is possible that they are the extremes of a continuum of movement that disperses all species of aphids.

Dispersal

Dispersal creates a relatively high mortality rate for sugarbeet root aphids. Additionally, alate aphids incur other disadvantages: if aphids fly then they may incur an additional cost in that their potential fecundity is further reduced and there is a further delay in the onset of reproduction. The combined effect is a marked reduction in reproductive potential and rate of increase. Although dispersal results in the colonization of suitable plants, it is not always clear what advantages there are in dispersing except from annual plants that are about to die. As dispersal can be costly in terms of fecundity or survival, or both, then sugarbeet root aphids are likely to delay departure until host quality falls below the average expectation for the habitat.

Behaviour

Life Cycle: In the spring, a female nymph (also known as a stem mother) emerges from an overwintering egg and initiates a gall on one of the leaves of the *Populus* tree. The gall forms around the stem mother, who begins to reproduce parthenogenically while feeding on the leaf's phloem sap. Each stem mother is capable of creating up to 300 progeny per gall. The gall occupants develop wings in the middle of the summer and disperse from the gall to deposit their larvae in the ground. These larvae colonize and feed on the roots of nearby *Chenopodiaceae* plants for the rest of the summer. In the summer, they form alate migrants that fly back to the *Populus* tree and asexually produce sexual males and females whose sole purpose is to mate, as they lack mouthparts to feed. The product of sexual reproduction is a single egg in each female which is deposited in the tree's bark and left to overwinter.

Leaf Colonization

Sugarbeet root aphids are closely synchronized with their hosts *Populus augustifolia*, with the majority of stem mothers colonizing leaves within three days after the leaf buds burst. There is intense competition between *Pemphigus* stem mothers over leaf choice - galls formed on larger leaves have higher stem mother weight, more aphids overall, and a lower probability of being aborted. Moreover, galls formed closer to the leaf stem-and thus closer to the source of nutrients flowing into the leaf-also benefit in the same way. During the aphids' emergence in the spring, large *Populus* leaves are colonized first. Once a stem mother forms a gall, she is more reluctant to move to a new leaf even if the current one withers, though the ability to colonize a large, healthy leaf close to its stem is crucial in ensuring an aphid's reproductive success.

On average, stem mothers distribute themselves among the leaves of a *Populus* tree according to the ideal free distribution model. Stem mothers sharing a leaf have to split the available resources, and this sharing comes at a cost to the stem mother's reproductive success. Two stem mothers sharing a leaf spend a significant amount of their time engaging in territorial behaviour instead of feeding or gall forming. Thus, stem mothers sharing a leaf produce less offspring than single stem mothers on leaves of the same size. When one stem mother is removed from a shared leaf, the reproductive success of the remaining occupant(s) rise accordingly. Some stem mothers choose to settle alone on smaller leaves instead of sharing a larger leaf with another individual. This creates a scenario where, on average, there is no difference between the reproductive success of stem mothers occupying leaves singly and stem mothers sharing leaves with other individuals.

As a result of the importance of leaf choice in their reproductive success, sugarbeet root aphid stem mothers are highly territorial and will compete with each other for the chance to form galls at the bases of the largest leaves. This competition usually takes the form of kicking and shoving contests; two stem mothers will align rear-to-rear and push against the other forcefully using their hind legs. The winner of these contests then settles closest to the leaf base, and the loser settles more distally. If the basal stem mother dies or is removed, the distal stem mother often moves down to the base of the gall to replace her.

These territorial contests are unusually long, and can span the course of several days. A side effect of this intense territorial competition is the creation of a "floater population" of unsettled stem mothers searching for unoccupied leaves.

Gall Formation

Stem Mothers and Gall Choice

Sugarbeet root aphid stem mothers induce gall formation in the leaf by probing the leaf tissue with their stylets. This leads to the formation of a small depression on the leaf, which eventually closes up over the stem mother and forms a gall. The extent of the probing activity dictates gall size, and removing the stem mother early on in the process leads to the formation of an unclosed, rudimentary gall. The extent of probing activity is correlated with the aphid's reproductive success. The probing activity may also serve a secondary purpose in helping the aphid determine leaf size. Leaf size is an important factor in determining what leaf a stem mother chooses to develop a gall on, but at the time of leaf colonization, leaves that are being colonized are only a tenth to a quarter of their mature size. It has been suggested that, during the probing of the leaf, the stem mother gauges the chemical environment of the leaf and uses that as a predictor of final leaf size and suitability as a gall site. Aphids possess chemosensory organs, allowing them to respond to a variety of stimuli from the plant, both those of the volatile variety (olfaction) and non-volatile variety (taste). Once the aphid is on a plant, it tests the surface of plant with its antennae and probes the plant with its mouth parts. The antennae contain many sensilla, allowing them to sense tactile and chemical stimuli.

Stem Mothers and Their Progeny

At the onset of autumn in temperate regions many species switch to the production of sexual forms, with each clone producing both egg-laying females and males. This switch is triggered by the longer nights in the autumn. Both sexual males and sexual females are produced parthenogenetically in response to external and/or internal cues, such as the amount of food present, the amount of daylight, and the quality of the leaf. However, sugarbeet root aphids are also capable of reproducing sexually. Aphids predominate in the temperate regions of the world; to overwinter in a cold-resistant resting stage, the fertilized egg, is an adaptation to temperate conditions. One possible short-term advantage of sex is that it generates siblings with a range of genotypes, and a range is more likely to include the fittest genotype for a particular patch than the single genotype of an asexual sibling-ship would. Thus, genetically diverse siblings could have more 'elbow-room' as they are potentially capable of exploiting more than one kind of leaf-patch.

Interaction with the Leaf

The benefits of settling basally are significant, with basal stem mothers producing 49-65% more offspring than their distally settled counterparts. The benefits of settling basally relate to the aphid's ability to manipulate the plant's food resources. The galls formed by sugarbeet root aphids act as physiologic sinks, diverting and intercepting the plant's normal transport of resources and nutrients. ^{14}C labelling experiments have shown that their galls intercept resources being transported from the midvein to the distal parts of the leaf. In addition, these galls are able to divert ^{14}C from neighbouring leaves. One study showed that, on average, 29% of the ^{14}C accumulating inside a basal gall was supplied by a neighbouring leaf and not the leaf the gall itself was on. In contrast, neighbouring leaves only supplied 7% of a distal gall's ^{14}C, illustrating the importance of settling basally.

Tree Colonization

After emerging in the autumn, the migrant forms of sugarbeet root aphids seek out *Populus* trees to colonize. These migrant forms are short-lived, and usually die within 12 to 48 hours. In selecting trees to colonize, the autumn migrants of sugarbeet root aphids prefer to colonize larger trees over smaller ones, and are likely to use simple cues such as tree crown size or tree resistance to colonization to decide on which trees to colonize. Despite the importance of leaf size to stem mothers, autumn migrants do not appear to take leaf size into consideration when choosing a tree to colonize.

Agricultural Impact

Sugarbeet roots have become a common crop for sucrose production in the northern United States. A concern among farmers is the impact that sugarbeet root aphids can have on these crops, which are colonized by and used as secondary hosts for sugarbeet root aphids.

Effect on Crops

The lifecycle of sugarbeet root aphids involves primary and secondary host plants. Galls are formed on the primary host, cottonwood, in the spring by the stem mother. Her wingless offspring, called apterae, feed on the gall until giving rise to winged individuals called alatae. These alatae break out of the gall and colonize the roots of their secondary hosts, sugarbeet.

These individuals then suck the sap from the sugarbeet roots, which causes them to lose their colour, wilt, and die. Infestation is apparent where they appear as circular patches in which plants and leaves are wilted and dying. During dry years when cracks form in the soil, the secondary host roots become much more accessible to the aphids, which can lead to severe yield loss.

Economic Costs

The economic impact of sugarbeet root aphids on sugarbeet crops in southwestern Minnesota was studied during the 1990 and 1991 growing seasons. The effects of infestation on yield and quality of sugarbeet root showed that loss of sucrose content in the plant was the primary reason for reduced quality. In addition, yield rates were significantly higher in 1991 due to higher levels of precipitation. Further studies have shown that sugarbeet root aphid infestation is most prevalent in the upper Midwest during drought years, and in the southwest during times of low irrigation. Sometimes, despite dry conditions, cool weather can decrease the prevalence of sugarbeet root aphid infestations.

Pest Control and Management

Although certain control methods are effective on other root-feeding arthropods, control measures for sugarbeet root aphids are more difficult. Crop rotation and simple foliar and postemergence insecticides are usually ineffective. However, Knox Out 2FM and Counter 15G have proven to be effective in containing infestations, although some states, such as California, currently have no chemicals registered for use on the sugarbeet root aphid. Biological controls, such as introduction of fungal diseases or natural predators, may also serve as an effective means of pest control. Although it is unlikely that biological controls are fully capable of controlling sugarbeet root aphid populations, future research may increase their role in management.

To properly manage damaged and infested areas, it is necessary to thoroughly work these areas and destroy plants left in the ground for the following harvest. Weeds in the infested areas should be destroyed, equipment should be cleaned, and infested fields should not be used for at least three years. Water stresses should be avoided in order to prevent yield loss due to water-stressed sugarbeets.

Overwintering Biology of the Sugarbeet Root Aphid: Development & Validation of a Forecasting Model

Pemphigus betae Doane, occurs throughout the major sugarbeet growing areas of North America, including infestations in Texas (Winter

1980), Alberta, Canada (Harper 1963), California (Grigarick & Lange 1962) and Michigan (Landis 1990, personal communication). SRA infestations have nearly always been more severe under dry soil moisture conditions, due either to dry years under dry land conditions in the upper midwest, or to using less water in the irrigated areas of the west and southwestern U.S.

In Minnesota, the drought of the late 1980's, especially during 1988 and 1989, was probably most responsible for the buildup of economically damaging SRA infestations in the southwestern part of the state. In contrast, 1990 and 1991 were very wet years with soil moisture returning to normal levels in southern Minnesota. Although the 1992 season was dryer, it was one of the coolest seasons on record. Not too surprisingly, SRA infestations were less abundant during the past three seasons and infestation "hot spots" developed much later compared to the previous drought years (about mid- to late-Aug. vs. July 1st).

Of the studies that have been done to determine the impact of SRA on yield, there have been conflicting results with regard to actual sugar losses. For example, Maxson (1916) observed a 4.55 T/A (25% less than uninfested fields) loss in beet yield with sugar reductions of <1%. Harper (1961) reported yield losses of 1.3 to 3.1 T/A but did not record sugar losses. Most recently, Summers and Newton (1989) found that SRA in California reduced yield by 36-60%, but in only one of four commercial fields studied, did they observe the sugar percentage to be significantly lower in infested areas. However, because of the drastic reduction in beet yield, total sugar/A was significantly reduced in three of four fields. Our results from the 1990-1991 seasons in Minnesota clearly indicate that if the infestation approaches a beet rating of > 2.75, beet yields are typically reduced by 50% and sugar percentage by 40-50% within aphid hot-spots. While gross yields were variable, sugar content and actual recoverable sugar per acre in SRA infested areas were significantly ($P<0.05$) reduced in 73% (8/11) of the fields examined (Hutchison & Campbell 1992,1993).

Because of the sporadic and cryptic nature of subterranean SRA, very little work has been done to understand the biology, ecology or management of the pest on sugarbeet. In addition, damaging infestations continue to show up locally each year. If unusually dry conditions occur in subsequent years, more widespread damaging infestations are likely as the commercial hybrids grown in Minnesota are highly susceptible to SRA. Without a basic knowledge of the life cycle of SRA and its mechanisms of infestation, high risk fields cannot be identified nor can management strategies be effectively implemented. The focus of this article is to provide a summary of the work we have done on overwintering and spring phenology of SRA, validation of a model to forecast spring phenology, and summer population dynamics. This information will be useful for the development of a risk-rating system for estimating which fields may be likely to become infested.

Overwintering Biology

In many locations, SRA's primary host is narrow-leaf cottonwood, which probably does not occur in Minnesota. Narrow-leaf cottonwood usually occurs only at elevations >5,000 ft. In addition, the aphids we have collected from galls on our native "eastern" cottonwood, were not identified as P. *betae,* but as other related *Pemphigus* species. Despite this finding, we have observed significant numbers of a late (winged) aphids being produced in the fall. After emerging from sugarbeets, aphids have been trapped on yellow sticky cards placed on nearby cottonwood. However, trees in the area were monitored closely during the spring of 1991 for galls; no SRA developed on these trees. We also exposed SRA alates emerging from potted sugarbeet plants in the greenhouse to cottonwood bark to determine whether or not SRA will lay eggs for overwintering. Alates collected from field samples and potted beets were dissected and found to contain an average of 6.5 sexuales/female. However, these sexuales did not lay the eggs that overwinter on cottonwood trees.

The most obvious SRA overwintering strategy appears to be adult (non-winged) survival in the soil. Figure depicted earlier, summarizes our current model of SRA overwintering in Minnesota. Examples of SRA movement from overwintering sites to beets are shown in Figure earlier, where the SRA dispersed from culled beets in 1991, or fallow soil in 1992, to lambsquarter. Similar results from both years' research clearly demonstrates the role of lambsquarter in facilitating the spring buildup and movement of aphids into beet fields. The orientation of the 1992 infestations, in relation to the adjacent 1991 hot spots, are illustrated in Figure. These results also suggest that good weed management can help to suppress SRA infestations. The decline of SRA on lambsquarter in late June also corresponds to the time at which we can begin to see aphids building up in newly planted beet fields. An additional piece to the puzzle is that first-instar nymphs (immature stage), rather than adults, could very well be the primary migrating stage. We have observed them moving in wind currents in large incubators in the laboratory.

Figures depicted earlier, summarize the life table (death and birth rates) studies generated for overwintering SRA adults collected in

March-April, 1992. This is the first publication of age-specific life tables for the a non-feeding adult stage in aphids. A lower threshold for development was estimated from the developmental time (collection date to first reproduction) data to analyze all temperature data on a degree-day (DD) time scale. As with crop development, degree-days are a useful tool for quantifying the impact of temperature on insect reproduction, development, etc. Using the 6.6°C threshold, time to first reproduction averaged from 90-100 DDs. This DD estimate incorporated fluctuating temperature accumulations in the field up to the collection date, as well as constant temperatures in the laboratory. As illustrated, despite the different collection dates, and rearing temperatures, reproduction and survival patterns are quite similar for each of the three constant temperatures, with most of the reproduction occurring between 100-200 DDs. On a day (calendar) time scale, however, SRA at 12°C took nearly 3X longer to complete reproduction (mx) than at 24°C.

The degree-day (DD) model is also in agreement with an independent field observation of the progression of aphids that are reproductive (i.e., containing live embryos, but not yet larvipositing).

These studies are on-going through the 1992-1993 winter season and will continue into the spring of 1993 and winter of 1993-1994. Based on more conclusive evidence from the 1991-1992 winter and 1992 fall, we are still processing field samples that should enable us to document the proportion of aphids that overwinter in the soil and those that produce winged forms in the fall, that is an apparent latent, suicidal behaviour in Minnesota. To do these studies, we are again monitoring a small plot (100 rows X 250 ft) of sugarbeets planted in 1992 where we had confirmed SRA "hot-spots" during the season (Clara City location). In 1993 we will again be monitoring any possible gall formation on cottonwood trees adjacent to the sugarbeet plot to refute the possibility of gall forming aphids infesting beet fields.

In addition to documenting field colonization in the spring, a primary goal of the 1993 season will be the validation of the degree-day model at a minimum of four field sites. At selected sites, we will set up ambient air and soil temperature recording probes for site specific temperature data. In addition each site will be selected in close proximity to weather monitoring sites maintained by SMSC and for the University of Minnesota, to evaluate regional differences in temperature accumulation and the effect on forecasting precision. These data will provide a good foundation for testing the model in 1993. We see this step as an integral component of our overall goal to develop the early-warning system.

Spring & Summer Phenology

As part of the beet sampling project, an "end-of-season" area-wide survey has been conducted during each of the last three seasons. Of the 51 fields sampled in 1990, 11 had at least some SRA infestations (22 %), while only 5 were heavily infested (10 %). In 1991, the region-wide survey indicated that 9 of 39 fields (23 %) were infested with SRA. In 1992, 9 of 42 fields (21 %) were found to have one or more SRA infestations. These data will provide a good base-line for comparison with future years.

Within-Field Dispersal: As illustrated in Figures, during the cool 1992 season, SRA progressed slowly from field edges toward the centre of the field. The two figures summarize data from two different field edges in the same field. The primary difference was that the low site was adjacent to a 1991 beet field that was heavily infested, separated only by about 3' of grassy border. The Northeast corner infestation began with only a few infested beets along the field edge. The limited infestation along the edge of the northeast spot apparently limited the magnitude of the subsequent spread, relative to that of the low spot.

The detailed data collected in 1992 on field colonization will be very useful in validating the degree-day model of aphid spread during the season. Additional field sites, under different temperature/precip. Conditions, will be essential to complete model validation. Fields found to be SRA-infested in 1992 will be scouted in early 1993 to determine the extent to which aphids have successfully overwintered. New beet fields adjacent to or in close proximity of the 1992 SRA infestations will be sampled in 1993 to determine if and when migration and colonization has occurred. Early season scouting involves digging soil samples in areas where culled beets remain as well as examining roots of weed species known to host SRA (i.e. *Chenopodium* spp.). Early sampling is usually done on field margins and near field entrances. Any new infestations found will be monitored throughout the 1993 season to document the extent and rate of spread within the field. This was done in 1992 by weekly sampling of beets, row by row, within a 100 ft radius of the initial infestation. Soil samples taken from beets found to be infested provide estimates of aphid density and rate of increase.

Insecticides

During 1992, an insecticide plot was set up within each of the two dispersal sites at the Clara City location. Plots were arranged in a latin square design, with 6 treatments (20' x 3 rows) replicated 6 times. An untreated row separated each treatment. Treatments consisted of two rates of Knox Out 2FM as a directed spray and Counter 15G applied to the base of plants. Sprays were either incorporated manually or left as a lay-by. Granular treatments were incorporated manually. Treatments were applied when sampling indicated dispersal had occurred to at least 10 rows from the field edge.

Infestations did not spread throughout either of the plots. Therefore, only treatments within the first 12 rows were analyzed, and the two plots were combined to keep the number of replications at 6. Analyses based on percent infested, mean root rating, and yield showed no statistical significance.

These results are consistent with past insecticide trials involving SRA (Campbell & Hutchison 1991, 1992). However, aphid density samples for the 1992 test have yet to be counted, and may provide additional information. The data taken at the time of evaluation is based on the presence of wax produced by the aphids. This wax remains even after aphids have died or left the beet.

Therefore, percent infestation and root rating give no indication of whether viable colonies remained in treated areas. The information here indicates that infestations at the time of treatment were not at economically damaging levels, based on final root ratings (Hutchison & Campbell 1992). Levels of infestation in the untreated check remained below levels known to be economically damaging up to the harvest. This was most likely due to cool summer temperatures which hindered colony development and dispersal.

We will continue to conduct insecticide trials, using different materials and application timing strategies. As in 1992, we will attempt to conduct trials in at least three locations. We will also continue to focus on Counter 15G and Diazinon formulations (14G or Knox-Out), as these materials are currently labelled for sugarbeet. In a field situation, a decision to treat a large area with insecticides, would have to be made before the canopy closes. Early evaluations of SRA dispersal into the field, as well as knowing the likelihood of hot, dry conditions will be important in the decision making process. Economically damaging infestations may have to be present early in July before insecticides can be recommended.

All data gathered to this point indicates that insecticides are unpredictable in light pressure situations. Additional studies are needed to determine if early treatments can suppress dispersal. Given the fact that insecticidal control currently represents the only realistic option for SRA control, these studies are necessary and complementary to the degree-day forecasting research.

Sra Density - Damage - Weld Relationships

Examples of previous SRA-yield loss relationship were presented at previous research reporting sessions (Hutchison & Campbell 1991, 1992). As in 1990-1991 seasons, work was done to quantify SRA impact on sugar yields in commercial fields that were infested with SRA. A good cross-section of fields were sampled during the 1990-1991 field seasons. During the 1992 season, however, infested "hot spots" that were sampled (9/46 fields surveyed were infested), root ratings were < 2.5, indicating significant yield losses were unlikely.

These studies will be repeated in 1993 to provide data under different agronomic and environmental conditions (e.g., a warmer, dryer year). The same approach will be used. Five sets of 10-row-feet samples will be taken within SRA-infested "hot-spots" and compared to adjacent uninfested areas within the same field. As in 1990 and 1991, beet and sugar yield, as well as impurity analyses will be determined at the SMSC research laboratory. Beet root ratings and percentage of beets infested will also be determined for each yield sample. Where possible, we will also attempt to identify potential "hot-spots" relatively early (July-August) so that within-season estimates of SRA infestations can also be correlated with yield effects.

Army Cutworm (Euxoa Auxiliaris Grote)

The army cutworm (*Euxoa auxiliaris*) is a species of moth. Its nickname "miller moth" comes from the fine scales on its wings that rub off easily and remind people of the dusty flour that covers the clothing of a miller.

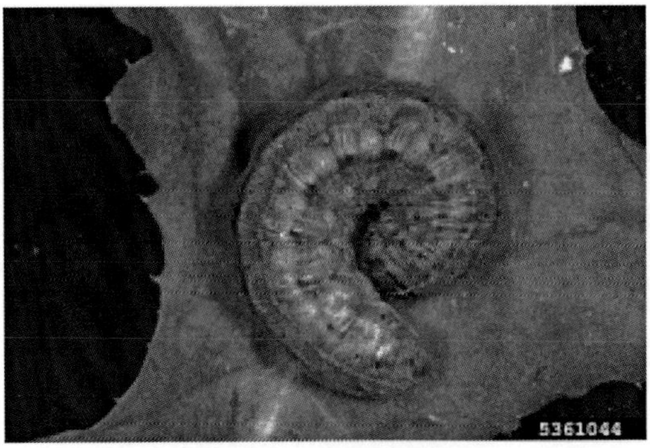

Figure: *Caterpillar*

Its caterpillars are pests of oat (*Avena sativa*) and common wheat (*Triticum aestivum*). The miller moth is a seasonal nuisance in the spring in states including Colorado, Wyoming, New Mexico and Kansas, as they hatch in the low-lying farmlands, then migrate to higher elevations for the summer. They return as the weather cools but in smaller numbers. They are considered nearly impossible to control through normal pest extermination techniques because a new batch shows up every day as they migrate. With their very small bodies, they enter homes (attracted by the light) in the evening through any available crack or crevice (door jamb, chimney, etc.) and residents in migration paths report dozens of moths per day entering their homes and garages. However, other than being a nuisance, they are not considered harmful.

Location

Euxoa auxiliaris is commonly found in the Western section and prairies of the United States. They are known to travel to alpine climate regions in late June and early July where they feed at night on the nectar of wildflowers. Army cutworms are one of the richest foods for predators, such as brown bears, in this ecosystem, where up to 72 per cent of the moth's body weight is fat, thus making it more calorie-rich than elk or deer.

Insect Life Cycle

Host Plants: Army cutworm larvae eat the foliage of wheat, oats, barley, mustard, flax, alfalfa, sweetclover, peas, cabbage, sugar beet, various weeds (notably stinkweed) and grasses.

Overwintering: The female moths each lay about 1,000 eggs in soft soil in late August through October. The eggs hatch in a few days to two weeks. The larvae feed above ground on plant foliage at night and remain below ground during the day. Development stops when the ground freezes; larvae are usually about half-grown by this time. They remain inactive throughout the winter just beneath the surface in loose soil.

Appearance Times: Larvae begin to feed in April and continue to feed until pupation in May to early June. Moths appear in June for a brief flight period, aestivate (summer hibernation) in buildings and under trash and clods during June and July, and then become active again for the egg-laying period.

Number of Generations: There is one generation per year.

Natural Enemies: Five or possibly six species of wasp parasites have been recorded for army cutworm in Canada, mostly from Alberta.

Copidosoma is a tiny parasitic wasp (about 2 mm in length) which lays a single egg in cutworm larvae. The egg produces multiple embryos from which over a thousand offspring may be produced. Sixty per cent or more of army cutworm larvae were parasitized by this organism during an outbreak in Alberta in 1990. Copidosoma prolonged the larval stage of army cutworm into June; the cutworms caused damage to spring crops, an unusual circumstance since they normally have pupated by that time.

Damage Assessment

Damage Description: Damage can be of any severity up to complete defoliation. In severe infestations, the area defoliated has ranged from an individual field to thousands of acres with larval densities of up to 200 per square metre. The first signs of damage are holes in leaves and semi-circular notches eaten from the edges of leaves.

Sampling and Monitoring Methods: Army cutworm moths have been monitored in southern Alberta with pheromone traps since 1978. Sample larvae as for other subterranean cutworm species. Mark an area of soil 50 cm by 50 cm. During the day, larvae should be within the top 5-7 cm of soil surface. Count the larvae within each 50 cm of row in the sample area. Multiply the number of larvae by four to give the number of larvae per square metre. Repeat the process in different areas of the field.

Economic Thresholds: Economic thresholds for this insect have not been rigorously tested. However, the following guidelines may be helpful.

The economic threshold is dependent, among other factors, on the crop infested. Mustard is more susceptible to damage from this cutworm than are cereals and alfalfa. A density of five cutworms per square metre is sufficient to destroy a mustard field, whereas cereals and alfalfa can withstand cutworm populations of up to 50 per square metre. The latter crops can resume growth after attack whereas mustard cannot.

Other factors affecting the economic threshold are the plants' growth condition and the number of weed hosts in the field. A weedy field will suffer less damage than a clean one. Forage crops and pastures must be watched closely in April and early May for the presence of these larvae. Plants that have adequate moisture and are vigorously growing with 12 to 15 cm of top growth can withstand four larvae per 30 cm of row without loss of yield. If plants are under 10 cm in height and two or more larvac per 30 cm of row are present, chemical treatment is required.

Management Strategy

Effects of Weather: Each outbreak year is usually preceded by a year with an abnormally dry July and wet autumn. A July with less than 3.8 cm of rainfall and a mean temperature of 17 degrees Celsius or higher is favourable to a population increase. However, an outbreak may not occur in the following year unless a dry July is followed by a total of over 11.4 cm of precipitation in August to October, with most of it in September. Army cutworms are reduced in a wet July when moths are drowned or covered with mud during aestivation. A dry fall delays egg-hatching by up to two months; eggs and first instar larvae are very susceptible to desiccation and are killed in the soil. Because of these mortality factors, an abundance of moths in early fall does not necessarily mean a cutworm outbreak the following year.

Cultural Practices: Spring crops can escape damage if they emerge after the cutworms have pupated. This is the usual situation and occurs when cutworm development is advanced by favourable fall or spring weather or crop seeding is delayed by wet weather. Crops can also be seeded later in the season to avoid attack by this pest.

Mechanical: If cutworms are marching, plow a steep-sided trench across the path of advance. Line the trench with a plastic sheet to ensure that larvae will not climb out of the trap. Use of insecticides on larvae in this limited space is economical.

Beet Armyworm (Spodoptera Exigua)

The Beet Armyworm or Small Mottled Willow Moth (*Spodoptera exigua*) is one of the best-known agricultural pest insects. It is also known as the asparagus fern caterpillar, and the adult moth is known in the British Isles (where it is an introduced species and not known to breed) as the small mottled willow. It is native to Asia, but has been introduced worldwide and is now found almost anywhere its many host crops are grown. The voracious larvae are the main culprits. They are greenish-brown cutworms, soft and bulging caterpillars with dark longitudinal stripes. The adult is a drab brown or grey moth 2 to 3 cm in wingspan.

Figure: *Larva*

The larvae feed on the foliage of plants, and can completely defoliate small ones. Smaller larvae devour the parenchyma of leaves, so that all that remains is the thin epidermis and veins. Larger larvae tend to burrow holes through thick areas of plants. For example, they will burrow straight into a head of lettuce rather than neatly removing tissue from one particular leaf. This renders the produce unmarketable. They attack buds and new growth on plants, preventing flowers from opening, new leaves from sprouting, and vegetables from developing. As the smaller larvae move about they leave strands of silk behind, netting the leaves with a silvery film.

The wide host range of the beet armyworm includes asparagus, beans and peas, sugar and table beets, celery, cole crops, lettuce, potato, tomato, cotton, cereals, oilseeds, tobacco, many flowers, and a multitude of weed species. . The beet armyworm does not tolerate cold. It can overwinter in warm areas, such as Florida and Hawaii, but in colder areas it dies off during the winter and the region is reinvaded by the adult moth as the weather warms and crop plants sprout.

Pupa: Pupation occurs in the soil. The chamber is constructed from sand and soil particles held together with an oral secretion that hardens when it dries. The pupa is light brown in colour and measures about 15 to 20 mm in length. Duration of the pupal stage is six to seven days during warm weather.

Adult: The moths are moderately sized, the wing span measuring 25 to 30 mm. The forewings are mottled gray and brown, and normally with an irregular banding pattern and a light coloured bean-shaped spot. The hind wings are a more uniform gray or white colour, and trimmed with a dark line at the margin. Mating occurs soon after emergence of the moths, and oviposition begins within two to three days. Oviposition extends over a three to seven day period, and the moths usually perish within nine to 10 days of emergence. Heppner provides a key to the adults of North American *Spodoptera*.

Host Plants

The beet armyworm has a wide host range, occurring as a serious pest of vegetable, field, and flower crops. Among susceptible vegetable crops are asparagus, bean, beet, broccoli, cabbage, cauliflower, celery, chickpea, corn, cowpea, eggplant, lettuce, onion, pea, pepper, potato, radish, spinach, sweet potato, tomato, and turnip. Field crops damaged include alfalfa, corn, cotton, peanut, safflower, sorghum, soybean, sugarbeet, and tobacco. Weeds also are suitable for larval development, including such common plants as lambsquarters, *Chenopodium album*;

mullein, *Verbascum* sp.; pigweed, *Amaranthus* spp.; purslane, *Portulaca* spp.; Russian thistle, *Salsola kali*; parthenium, *Parthenium* sp.; and tidestromia, *Tidestromia* sp.

Damage

Larvae feed on both foliage and fruit. In Florida it is regarded as a serious defoliator of flower crops and cotton, though much of the injury is induced by insecticide use that interferes with natural enemy activity. Young larvae feed gregariously and skeletonize foliage. As they mature, larvae become solitary and eat large irregular holes in foliage. They also burrow into the crown or centre of the head on lettuce, or on the buds of cole crops. As a leaf feeder, beet armyworm consumes much more cabbage tissue than the diamondback moth, *Plutella xylostella* (Linnaeus), but is less damaging than the cabbage looper, *Trichoplusia ni* (Hübner).

Tomato fruit is most susceptible to injury, especially near fruit maturity, but beet armyworm is not considered to be as threatening to tomato as is the corn earworm, *Helicoverpa zea* (Boddie).

Sampling

Pheromone traps can be used to detect the presence of adult beet armyworm. Visual sampling for damage and larvae, combined with an action threshold of 0.3 larvae per plant, was used successfully on cabbage in south Texas to determine the need for crop treatment with insecticides. Regular monitoring of crops, probably about twice per week, is recommended because adults frequently invade from surrounding crops or weeds.

Natural Enemies

Numerous native natural enemies have adapted to this pest. Among the most common parasitoids are *Chelonus insularis* Cresson, *Cotesia marginiventris* (Cresson), and *Meteorus autographae* (Muesbeck) (all Hymenoptera: Braconidae), and the tachinid *Lespesia archippivora* (Riley) (Diptera: Tachinidae). Predators frequently attack the eggs and small larvae; among the most important are minute pirate bugs, *Orius* spp. (Hemiptera: Anthocoridae); bigeye bugs, *Geocoris* spp. (Hemiptera: Lygaeidae); damsel bugs, *Nabis* spp. (Hemiptera: Nabidae); and a predatory shield bug, *Podisus maculiventris* (Say). Pupae are subject to attack, especially by the red imported fire ant, *Solenopsis invicta* Buren. Fungal diseases, *Erynia* sp. and *Nomurea rileyi*, and a nuclear polyhedrosis virus also inflict some mortality. The important mortality factors vary among crops, and among geographic regions. None except the nuclear polyhedrosis virus are highly specific to beet armyworm, which may explain why they are not especially effective. Virus is considered to be the most important mortality factor.

Insecticides

In the Southeast and Southwest, the relatively high abundance of beet armyworm has stimulated frequent application of insecticides to foliage. Insecticide resistance is a major problem in management of this insect, possibly because it attacks crops such as flowers, cotton, and vegetables - crops that are treated frequently with insecticides. Beet armyworm abundance is favoured by frequent insecticide use, and it is considered to be a secondary or induced pest in some crops. Also, intensive use of insecticides for beet armyworm control in vegetables such as celery has stimulated outbreaks of other pests, principally American serpentine leafminer, *Liriomyza trifolii* (Burgess).

Beet armyworm larvae are susceptible to neem products. Eggs can be killed with petroleum oil, and both eggs and young larvae can be controlled with foliar applications of 5% cottonseed oil, but this concentration is damaging to some plants.

Pheromones can also be used to disrupt mating and inhibit or eliminate reproduction. Saturation of the atmosphere around beet armyworm-susceptible crops has been estimated to reduce mating by 97%.

Biological Control

Several insect pathogens may prove to be useful for suppression of beet armyworm. A nuclear polyhedrosis virus isolated from beet armyworm is fairly effective as a bioinsecticide under greenhouse conditions, where inactivation by ultraviolet light in sunlight is not a severe problem. The fungus *Beauveria bassiana* has the same attributes and limitations. Entomopathogenic nematodes (Rhabditida: Steinernematidae and Heterorhabditidae) successfully infect both larvae and adults of beet armyworm.

Black Cutworm (Agrotis Ipsilon Hufnagel)

The black cutworm, *Agrotis ipsilon* (Hufnagel), has a wide host range, feeding on nearly all vegetables and many important grains.

Distribution

The origin of black cutworm is uncertain, though it is now found in many regions of the world, being absent principally from some tropical regions and cold areas. It is more widespread, and damaging, in the northern hemisphere than the southern hemisphere. It annually reinvades temperate areas, overwintering in warmer or subtropical regions. Long distance dispersal of adults has long been suspected in Europe, China, and North America.

The basic pattern is to move north in the spring, and south in the autumn. Studies in the United States demonstrated northward displacement of moths during the spring in the range of 1000 km in two to four days when assisted by northward flowing wind. Similar displacement to the south and southwest has been documented in the autumn.

Description and Life History

The number of generations occurring annually varies with weather conditions. In North America, there are one to two generations in Canada but two to four in the United States. In Tennessee, USA, moths are present in March-May, June-July, July-August, and September-December. Based on light trap collections, moths are reported to be abundant in Arkansas, USA (a warm climate) during May-June and September-October, and in New York, USA (a cool climate), they occur mostly in June-July. However, light traps are not very effective during the spring flight, and underestimate early season. Thus, the phenology of black cutworm remains uncertain, or perhaps is inherently variable due to the vagaries associated with long range dispersal.

Overwintering has been reported to occur in the pupal stage in most areas where overwintering occurs, but larvae persist throughout the winter in Florida, USA, a subtropical environment. Pupae have been known to overwinter as far north as Tennessee, but apparently are incapable of surviving farther north. Thus, moths collected in the central region of USA in March and April are principally dispersing individuals that are past their peak egg production period. Nonetheless, they inoculate the area and allow production of additional generations, including moths that disperse north into Canada. Duration of the life cycle is normally 35 to 60 days.

Adult: The adult is fairly large in size, with a wingspan of 40 to 55 mm. The forewing, especially the proximal two-thirds, is uniformly dark brown. The distal area is marked with a lighter irregular band, and a small but distinct black dash extends distally from the bean-shaped wing spot. The hind wings are whitish to gray, and the veins marked with darker scales. The adult preoviposition period is about seven to 10 days. Moths select low-growing broadleaf plants preferentially for oviposition, but lacking these will deposit eggs on dead plant material. Soil is an unsuitable oviposition site.

Figure: *Adult black cutworm,* Agrotis ipsilon *(Hufnagel). Photograph by John L. Capinera, University of Florida.*

Egg: The egg is white in colour initially, but turns brown with age. It measures 0.43 to 0.50 mm high and 0.51 to 0.58 mm wide and is nearly spherical in shape, with a slightly flattened base. The egg bears 35 to 40 ribs that radiate from the apex; the ribs are alternately long and short. The eggs normally are deposited in clusters on foliage. Females may deposit 1200 to 1900 eggs. Duration of the egg stage is three to six days.

Larva: There are five to nine instars, with a total of six to seven instars most common. Head capsule widths are about 0.26-0.35, 0.45-0.53, 0.61-0.72, 0.90-1.60, 2.1-2.8, 3.2-3.5, 3.6-4.3, and 3.7-4.1 mm for instars one through eight, respectively. Head capsule widths are very similar for instars one through four, but thereafter those individuals that display eight or nine instars show only small increments in width at each molt and eventually attain head capsule sizes no larger than those displaying only six or seven instars.

Larval body length is reported to be 3.5, 5.3-6.2, 7, 10, 20-30, 30-45, 50, and 50 mm for instars one through eight, respectively. Duration of the larval stage is normally 20 to 40 days. Mean duration of instars one through six was reported to be 6.0, 5.0, 4.6, 4.3, 5.6, 4.0 days, respectively, at 22°C. Larval development is strongly influenced by temperature, with the optimal temperature about 27°C. Humidity is less important, but instars one through five thrive best at higher humidities.

In appearance, the larva is rather uniformly coloured on the dorsal and lateral surfaces, ranging from light gray or gray-brown to nearly black. On some individuals, the dorsal region is slightly lighter or brownish in colour, but the larva lacks a distinct dorsal band. Ventrally, the larva tends to be lighter in colour. Close examination of the larval epidermis reveals that this species bears numerous dark, coarse granules over most of its body. The head is brownish with numerous dark spots. Larvae usually remain on the plant until the fourth instar, when they become photo-negative and hide in the soil during the daylight hours. In these latter instars they also tend to sever plants at the soil surface, pulling the plant tissue belowground. Larvae tend to be cannibalistic.

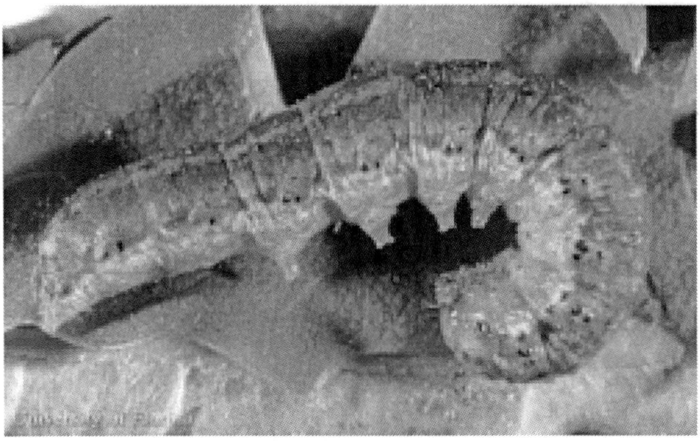

Figure: *Lateral view of the larva of a black cutworm,* Agrotis ipsilon *(Hufnagel).*

Pupa: Pupation occurs belowground at a depth of 3 to 12 cm. The pupa is 17 to 22 mm long and 5 to 6 mm wide, and dark brown. Duration of the pupal stage is normally 12 to 20 days.

Hosts Plants

Black cutworm has a wide host range. Nearly all vegetables can be consumed, and this species also feeds on alfalfa, clover, cotton, rice, sorghum, strawberry, sugarbeet, tobacco, and sometimes grains and grasses. In the midwestern USA it is considered to be a serious corn pest. Among the weeds suitable for larval development are bluegrass, *Poa pratensis*; curled dock, *Rumex crispus*; lambsquarters, *Chenopodium album*; yellow rocket, *Barbarea vulgaris*; and redroot pigweed, *Amaranthus retroflexus*. The preference by black cutworm for weeds is sometimes quite pronounced, and crops will be attacked only after the weeds are consumed. Adults feed on nectar from flowers. Deciduous trees and shrub such as linden, wild plum, crabapple, and lilac are especially attractive to moths.

Damage

This species occurs frequently in many crops, and is one of the best-known cutworms. Despite the frequency of occurrence, however, it tends not to appear in great abundance, as is known in some other cutworms and armyworms. Black cutworm is not considered to be a climbing cutworm, most of the feeding occurring at soil level. However, larvae will feed aboveground until about the fourth instar. Larvae can consume over 400 sq cm of foliage during their development, but over 80% occurs during the terminal instar, and about 10% in the instar immediately preceding the last.

Thus, little foliage loss occurs during the early stages of development. Once the fourth instar is attained, larvae can do considerable damage by severing young plants, and a larva may cut several plants in a single night. Plants tend to outgrow their susceptibility to injury. Corn at the one-leaf stage is very susceptible to damage, but that by the 4 or 5-leaf stage plant yield was not reduced by larval feeding. Leaf feeding and cutting above the soil line are less damaging to corn than cutting at the soil surface. Subterranean damage is very injurious.

Natural Enemies

Numerous species of natural enemies have been associated with black cutworm, but data on their relative importance are scarce. However, in Missouri, USA, there are reports of 69% parasitism, so natural enemies probably exact a significant toll on cutworm populations.

Among the wasps known to attack this cutworm are Apanteles marginiventris (Cresson), Microplitis feltiae Muesebeck, Microplitis kewleyi Muesebeck, Meteorus autographae Muesebeck, Meterorus leviventris (Wesmael) (all Hymenoptera: Braconidae); Campoletis argentifrons (Cresson), Campoletis flavicincta (Ashmead), Hyposoter annulipes (Cresson), and Ophion flavidus Brulle (all Hymenoptera: Ichneumonidae). Larvae parasitized by Meteorus leviventris (Wesmael) consume about 24% less foliage and cut about 36% fewer seedlings, so considerable benefit is derived from parasitism in addition to the eventual death of the host larva.

Other parasitoids known from black cutworm include flies often associated with other ground-dwelling noctuids, including Archytas cirphis Curran, Bonnetia comta (Fallen), Carcelia formosa (Aldrich and Webber), Chaetogaedia monticola (Bigot), Eucelatoria armigera (Coquillett), Euphorocera claripennis (Macquart), Gonia longipulvilli Tothill, G. sequax Williston, Lespesia archippivora (Riley), Madremyia saundersii (Williston), Sisyropa eudryae (Townsend), and Tachinomyia panaetius (Walker) (all Diptera: Tachinidae).

Predatory ground-dwelling insects such as ground beetles (Coleoptera: Carabidae) apparently consume numerous larvae.

Although studies in Florida, USA, indicated that 75 to 80% of cutworms could be killed by a granulosis virus, there is surprisingly little information on epidemiology and of natural pathogens. Rather, such pathogens as viruses, fungi, bacteria, and protozoa from other insects have been evaluated for black cutworm susceptibility; in most cases only relatively weak pathogens have been identified.

An entomopathogenic nematode, *Hexamermis arvalis* (Nematoda: Mermithidae), is known to parasitize up to 60% of larvae in the central USA. Entomopathogenic nematodes (Nematoda: Steinernematidae and Heterorhabditidae) will infect and kill black cutworm larvae, but their populations normally need to be supplemented to realise high levels of parasitism. Their effectiveness is related to soil moisture conditions.

Management

Survey and Trapping. Adult populations can be monitored with both blacklight and sex pheromone traps. However, light traps are not consistently efficient. Light traps are most effective in the summer and autumn, but the late season generations generally pose little threat to crops. Pheromone traps are more effective during the spring flight, when larvae present the greatest threat to young plants. Trap colour affects moth capture rate, with white and yellow traps capturing more than green traps.

Large larvae burrow in the soil, and are difficult to observe. However, larvae can be sampled with bait traps, and this is most effective prior to emergence or planting of seedlings. Various trap designs have been studied, but many employ a container sunk into the soil with the upper lip at the soil surface.

The container is baited with fresh plant material and/or bran, and with vermiculite so the larvae can attain shelter. Larvae are effectively captured in baited containers if the vermiculite is not too near the surface, and catches are enhanced if a screen cylinder, which provides a visual stimulus to the cutworms, is suspended above the baited container. If plants are present in the field they compete with the bait in the traps, and trap efficiency declines markedly. The distribution of larvae in the spring is random.

Chemical. Persistent insecticides are commonly applied to plants and soil for black cutworm suppression, but surface rather than subsurface soil applications are desirable. Larvae readily accept insecticide-treated bran and other baits Application of systemic insecticides to seeds also provides some protection against larval injury. *Bacillus thuringiensis* is not usually recommended for cutworm control.

Cultural and Mechanical. Black cutworm larvae feed readily on weeds, and destruction of weeds can force larvae to feed exclusively on crop plants, exacerbating damage.

Thus, it is often recommended that weeds not be tilled or treated with herbicide until larvae have matured. Timing is important, however, because prolonged competition between crop and weed plants can reduce crop yield. Presence of flowering weeds also can be beneficial by supporting prolonged survival of parasitoids. In contrast, reduced tillage cropping practices, which often produce higher weed populations, seem to result in increased abundance of black cutworm and higher levels of cutting in corn. This may be due, in part, to the tendency of moths to oviposit on weeds; weedy fields tend to have higher cutworm populations.

Black cutworm populations also tend to be higher in wet areas of fields, and in fields that have been flooded. Black cutworm has been known, at times, as "overflow worm," due to its tendency to be abundant and damaging in fields that have been flooded by overflowing rivers.

In the home garden, barriers are sometimes useful to prevent damage to seedlings by cutworms. Metal or waxed paper containers with both the top and bottom removed can be placed around the plant stem to deter consumption. Aluminum foil can be wrapped around the stem to achieve a similar effect. Because larvae will burrow and feed below the soil line it is necessary to extend to barrier below the soil surface. Because black cutworm moths, which easily circumvent such barriers, are active during the growing season, this procedure alone may have little value. Use of netting or row covers, in addition to larval barriers, should prove more effective.

Celery Looper (Anagrapha Falcifera)

Figure: *Adult on leaf, showing the great camouflage*

The Celery Looper (*Anagrapha falcifera*) is a moth of the Noctuidae family. It is found from Newfoundland, Labrador and Southern Canada to Georgia, Mississippi, Texas, Arizona, Idaho, Washington and Oregon.

The Wingspan is 35–40 mm.

A medium-size (3.5-4.0 cm wingspan) moth with two colour-forms, dark-brown or grey. The grey form, most common early in the season, is light grey with little mottling, shading to dark brown in the lower half of the forewing, between the antemedian and postmedian lines and in a band from the apex to the outer lower margin. The brown form is dark red-brown over most of the forewing except for a paler grey brown area at the wing base and along the outer margin. Stigma is a curved silver bar, rounded on the outer end, with a short spur on the upper arm. The lower arm continues as the lower half of the antemedian line. Hindwings in both forms are light red-brown, with a distinct narrow postmedian band and shading to a wide dark terminal band. The antennae are simple and boht the sexes are alike.

The larvae feed on beets, blueberries, clover, corn, lettuce, plantain, viburnum and other low plants.

15

Marigold and Pests

Calendula officinalis (pot marigold, ruddles, common marigold, garden marigold, English marigold, or Scottish marigold) is a plant in the genus *Calendula* of the family Asteraceae. It is probably native to southern Europe, though its long history of cultivation makes its precise origin unknown, and it may possibly be of garden origin. It is also widely naturalised further north in Europe (north to southern England) and elsewhere in warm temperate regions of the world.

Description

It is a short-lived aromatic herbaceous perennial, growing to 80 cm (31 in) tall, with sparsely branched lax or erect stems. The leaves are oblong-lanceolate, 5–17 cm (2–7 in) long, hairy on both sides, and with margins entire or occasionally waved or weakly toothed. The inflorescences are yellow, comprising a thick capitulum or flowerhead 4–7 cm diameter surrounded by two rows of hairy bracts; in the wild plant they have a single ring of ray florets surrounding the central disc florets. The disc florets are tubular and hermaphrodite, and generally of a more intense orange-yellow colour than the female, tridentate, peripheral ray florets. The flowers may appear all year long where conditions are suitable. The fruit is a thorny curved achene.

Cultivation

Calendula officinalis is widely cultivated and can be grown easily in sunny locations in most kinds of soils. Although perennial, it is commonly treated as an annual, particularly in colder regions where its winter survival is poor, or in hot summer locations where it also does not survive.

Figure: *A double-flowered cultivar*

Calendula are considered by many gardening experts as among the easiest and most versatile flowers to grow in a garden, especially since they tolerate most soils. In temperate climates, seeds are sown in spring for blooms that last throughout the summer and well into the fall. In areas of little winter freezing (USDA zones 8-11), seeds are sown in autumn for winter colour, plants will wither in subtropical summer. Seeds will germinate freely in sunny or half-sunny locations, but plants do best if planted in sunny locations with rich, well-drained soil. Pot marigolds typically bloom quickly from seed (in under two months) in bright yellows, golds, and oranges.

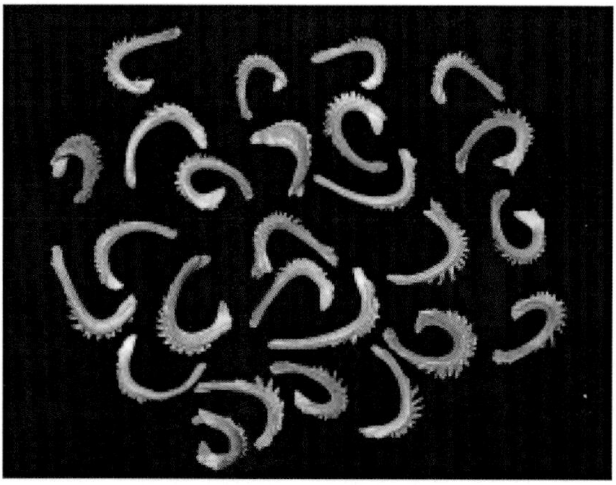

Figure: *Seeds*

Leaves are spirally arranged, 5–18 cm long, simple, and slightly hairy. The flower heads range from pastel yellow to deep orange, and are 3–7 cm across, with both ray florets and disc florets. Most cultivars have a spicy aroma. It is recommended to deadhead (removal of dying flower heads) the plants regularly to maintain even blossom production.

Numerous cultivars have been selected for variation in the flowers, from pale yellow to orange-red, and with 'double' flowerheads with ray florets replacing some or all of the disc florets. Examples include 'Alpha' (deep orange), 'Jane Harmony', 'Sun Glow' (bright yellow), 'Lemon' (pale yellow), 'Orange Prince' (orange), 'Indian Prince' (dark orange-red), 'Pink Surprise' (double, with inner florets darker than outer florets) and 'Chrysantha' (yellow, double). 'Variegata' is a cultivar with yellow variegated leaves.

Calendula are used as food plants by the larvae of some Lepidoptera species including Cabbage Moth, The Gothic, Large Yellow Underwing and Setaceous Hebrew Character.

Uses

Pot marigold florets are considered edible. They are often used to add colour to salads, or added to dishes as a garnish and in lieu of saffron. The leaves are edible but are often not palatable. They have a history of use as a potherb and in salads.

Flowers were used in ancient Greek, Roman, Middle Eastern and Indian cultures as a medicinal herb as well as a dye for fabrics, foods and cosmetics. Many of these uses persist today. They are also used to make oil that protects the skin.

Constituents

The petals and pollen of Calendula officinalis contain triterpenoid esters and the carotenoids flavoxanthin and auroxanthin (antioxidants, and the source of the yellow-orange colouration). The leaves and stems contain other carotenoids, mostly lutein (80%) and zeaxanthin (5%), and beta-carotene. Plant extracts are also widely used by cosmetics, presumably due to presence of compounds such as saponins, resins and essential oils.

The flowers of Calendula officinalis contain flavonol glycosides, triterpene oligoglycosides, oleanane-type triterpene glycosides, saponins, and a sesquiterpene glucoside.

Why Plant Marigolds and Other Plants in Vegetable Garden?

Marigolds are easy to grow and they help keep the away aphids. "The relationship between plants and insects is known as 'companion planting.' it's by far the safest, natural way to garden organically."

Plants That Naturally Repel Insects: There are many beneficial herbs that keep insects away.

- Peppermint repels ants, white cabbage moths, aphids, and flea beetles.
- Garlic discourages aphids, fleas, Japanese beetles, and spider mites.
- Perennial Chives repel aphids and spider mites.
- Chives are often planted among roses to keep aphids away and to resist the disease, Blackspot.
- Basil drives away flies and mosquitoes.
- Borage deters that monster of vegetable garden insects, the tomato hornworm.
- Rosemary and Sage repel cabbage moths, bean beetles, and carrot flies.
- Annual Marigolds can be used anywhere to deter Mexican bean beetles, squash bugs, thrips, tomato hornworms, and whiteflies. They are also known to repel harmful root knot nematodes (soil dwelling microscopic white worms) that attack tomatoes, potatoes, roses, and strawberries. The root of the Marigold produces a chemical that kills nematodes as they enter the soil. If a whole area is infested, at the end of the season, turn the Marigolds under so the roots will decay in the soil. You can safely plant there again the following spring.
- Nasturtium is another annual, in this case a trailing vine, that keeps away Colorado potato bugs, squash bugs, and whiteflies.
- The perennial, Artemisia or Wormwood, deters slugs that are so devastating to foliage.
- Radishes can be planted to discourage cucumber beetles, squash bugs, and stink bugs.

Attracting Good Insects to Your Garden:

Plant certain vegetables, herbs, or flowers in your garden to attract predatory insects that will feed on the harmful, undesirable ones.

Perennial Yarrow...attracts ladybugs that consume masses of aphids. The lacewing that feeds on aphids, mealy bugs, mites, and scale needs lots of pollen from flowers and evergreens for shelter. Wasps and bees are also beneficial to the garden. Even the prehistoric-looking preying mantis is a friend, so don't discourage it from visiting. When you create a natural balance in your garden you'll discover how much better everything grows and you won't need to worry about damaging the environment. They also suggest.

Ideal Planting Companions for Vegetables: The following is a list of vegetables and their ideal planting companions, plus combinations to avoid:

- Beans-like celery and cucumbers but dislike onions and fennel.
- Beets are compatible with bush beans, lettuce, onions, kohlrabi, and most members of the cabbage family. Keep pole beans and mustard away from them.
- Cabbage, celery, dill, onions, and potatoes are good companion plants. Dislikes include strawberries, tomatoes, and pole beans.
- Carrots, lettuce, radish, onions, and tomatoes are friends. Dill isn't, so plant it at the other end of the garden.
- Corn prefers to be near pumpkins, peas, beans, cucumbers, and potatoes. Keep tomatoes away.
- Cucumbers like sweet corn, peas, radishes, beans, and sunflowers. Dislikes include aromatic herbs and potatoes.
- Lettuce grows especially well with onions. They are also compatible with strawberries, carrots, radishes, and cucumbers.
- Onions can be planted near lettuce, beetroot, strawberries, and tomatoes but keep well away from peas and beans.
- Peas, carrots, cucumbers, sweet corn, turnips, radishes, beans, potatoes, and aromatic herbs are good companions. Keep peas away from onions, garlic, leek, and shallots.
- Radish grows well with beetroot, carrots, spinach, parsnip, cucumbers, and beans. Avoid planting near cabbage, cauliflower, Brussels sprouts, broccoli, or turnips.
- Squash can be planted with cucumbers and corn.
- Tomatoes, carrots, onions, and parsley are good companion plants. Basil improves growth and flavour. Keep cabbage and cauliflower away from them.

Gopher

Pocket gophers, commonly referred to as gophers, are burrowing rodents of the family Geomyidae. There are approximately 35 species of gopher living in Central and North America. They are commonly known for their extensive tunneling activities. Gophers are endemic to North and Central America.

The name "pocket gopher" on its own may be used to refer to any of a number of genera within the family. These are the "true" gophers; however there are several ground squirrels of the family Sciuridae that are often mistakenly called gophers as well.

Description

Gophers weigh around 0.5 pounds (230 g), and are about 6–8 inches (150–200 mm) long in body length, with a tail 1–2 inches (25–51 mm) long. A few species reach weights approaching 1 kg (2.2 lb). Within any species, the males are larger than the females and can be nearly double their weight.

Their lifespan is normally one to three years assuming no diseases or predation. The maximum life span for the pocket gopher is approximately five years. Some gophers, such as those in the genus geomys, have lifespans that have been documented as up to seven years in the wild.

Most gophers have brown fur that often closely matches the colour of the soil in which they live. Their most characteristic features are their large cheek pouches, from which the word "pocket" in their name derives. These pouches are fur-lined, and can be turned inside out. They extend from the side of the mouth well back onto the shoulders. They have small eyes and a short, hairy tail, which they use to feel around tunnels when they walk backwards.

Pocket gophers examined by humans have been found to often have external parasites on them. Common predators of the gopher include weasels, snakes and hawks.

Behaviour

All pocket gophers create a network of tunnel systems that provide protection and a means of collecting food. They are larder hoarders, and their cheek pouches are used for transporting food back to their burrows. Gophers can collect large hoards. Unlike its close relative the ground squirrel, gophers do not live in large communities and seldom find themselves above ground.

The entrances can be identified by small piles of loose dirt covering the opening. Their burrows can be found in many areas where the soil is softer and easily tunneled. They will often appear in vegetable gardens, lawns, or farms, as gophers like moist soil. This has led to their frequent treatment as pests.

Gophers eat earthworms, grubs, plant roots, shrubs and other vegetables such as carrots, lettuce, radishes, and any other vegetables with juice. Some species are considered agricultural pests. The resulting destruction of plant life will then leave the area a stretch of denuded dirt.

Pocket gophers are solitary outside of the breeding season, aggressively maintaining territories that vary in size depending on the resources available. Males and females may share some burrows and nesting chambers if their territories border each other, but in general, each pocket gopher inhabits its own individual tunnel system. Although they will attempt to flee when threatened, they may attack other animals, including cats and humans, and can inflict serious bites with their long, sharp teeth.

Depending on the species and local conditions, pocket gophers may have a specific annual breeding season, or may breed repeatedly through the year. Each litter typically consists of two to five young, although this may be much higher in some species. The young are born blind and helpless, and are weaned at around forty days.

Control

Geomys spp and Thomomys spp are classed as "prohibited new organisms" under New Zealand's Hazardous Substances and New Organisms Act 1996 preventing it from being imported into the country.

Classification

Figure: *A typical pocket gopher.*

There has been much debate among taxonomists about which races of pocket gopher should be recognised as full species, and the following list cannot be regarded as definitive.

Rodent Control

Rats and mice have long been a problem on farms where food and nesting sites are plentiful. These animals consume and contaminate food destined for livestock and other animals, as well as humans. Each rat on a farm will eat, spoil or damage approximately $25 worth of grain per year. The adaptability and agility of these animals make getting rid of them particularly difficult. Mice are capable of running up a vertical surface, negotiating a wire like the finest circus performer and can easily jump to a height of 30 cm (12 in.) from a flat surface.

Why Control Rodents?

Damage comes in many forms:

- Damage to buildings. Mice and rats will damage wood and electrical wiring, which can be a fire hazard.
- Destruction of insulation. Many livestock and poultry facilities show serious deterioration within 5 years. Associated with this damage are costs for re-insulation, increased energy costs and poorer feed conversions by animals.
- Feed consumed. A colony of 100 rats will consume over 1 tonne of feed in 1 year.
- Feed contaminated. A rat can contaminate ten times the amount of feed it eats with its droppings, urine and hair. A rat produces 25,000 droppings per year, a mouse 17,000. The U.S. Department of Agriculture estimates that the equivalent of more than $2 billion in feed is destroyed by rodents each year.
- Biosecurity. Rodents are recognised as carriers of approximately 45 diseases, including salmonellosis, pasteurellosis, leptospirosis, swine dysentery, trichinosis, toxoplasmosis and rabies. Mice and rats can carry disease-causing organisms on their feet, increasing the spread of disease.

Understanding Rodents

Mice and rats have tremendous breeding potential. Under ideal situations, a pair of rats and their offspring can produce 20,000,000 young in 3 years. Mice reproduce even faster. One female mouse can give birth to 5-10 litters per year, yielding five to six young per litter. The gestation period is a mere 19-21 days. These babies are sexually mature in 6-10 weeks. The average female mouse lives to be 9 months. One female rat is capable of producing another 22 breeding females in 1 year (assuming a 50:50 male/female ratio of offspring), which mature in 3 months after parturition.

Rats and mice have poor eyesight but excellent senses of smell, taste, touch and hearing. They do not like open areas and prefer contact with walls and other objects. They do not range far from the nest. The maximum range for rats is 45 m (148 ft), for mice 9 m (30 ft). Rats are extremely apprehensive about new objects and will avoid them for several days. Leaving a trap out for about 5 days is necessary to ensure acceptance. Mice quickly accept new objects. These tendencies become very important when designing baiting or trapping programs.

Does Your Farm have a Problem?

Producers should not be embarrassed to admit they have a rodent problem. Surveys in Ontario indicate that 80% of poultry producers and 89% of swine producers have rodent problems. The probability is that mice or rats currently exist on your farm. The embarrassment and costs occur if something is not done to confront the problem.

The following are signs of rodent infestation:

- Sounds. Gnawing, climbing noises in walls, squeaks.
- Droppings. Found along walls, behind objects and near food supplies.
- Burrows. Rat burrows are indicated by fresh diggings along foundations, through floorboards into wall spaces.
- Runs. Look for dust-free areas along walls and behind storage material.
- Gnawing marks. Look for wood chips around boards, bins and crates. Fresh gnawing marks will be pale in colour.
- Rodent odours. Persistent musky odours are a positive sign of infestation.
- Visual sighting. Daylight sighting of mice is common. Rats are seen in daylight only if populations are high. Quietly enter your barn at night, wait in silence for 5 minutes and listen for the sound of rodent activity. Look around with a powerful flashlight; rat eyes will reflect the light.
- Smudge marks. These may be found on pipes or rafters where dirt and oil from their fur leave a greasy film.

It is a generally accepted rule of thumb that there are approximately 25 mice or rats for every one that is seen.

Is It a Rat or a Mouse Problem?

Since rats and mice require different control strategies, determine whether the problem is rats or mice. The simplest way to differentiate between the types of infestation is by examining the droppings. Mouse droppings are black and rice-kernel size, whereas rat droppings are black and bean-sized.

What do Mice and Rats Like to Eat?

Rats and mice can be considered to be omnivorous. Given a choice, they prefer cereal grains. Rats eat meat when available. However, when food supplies are scarce, they will eat almost anything, including plaster and even soap or animal carcasses. Mice have been known to nest over winter inside the carcass of a deer stricken with rabies, consume the meat and become infected. They then become vectors of this disease. Rats and mice eat every day and prefer a water supply. Rats usually drink every day, but mice can survive several days without water.

Table: *Physical and behavioural characteristics of adult rats and mice*

Characteristic	***Norway Rat***	***Mouse***
Size (including tail)	42 cm (16.5 in.)	16 cm (6 in.)
Average weight (adult)	500 gm (18 oz)	20 gm (0.7 oz)
When active	nocturnal	nocturnal
Sight	poor (1.5 m) (4.9 ft)	poor (1 m) (3 ft)
Smell, touch, taste	excellent	excellent
Hearing	highly accurate	highly accurate
Range from nest	45 m (148 ft)	9 m (30 ft)
Fear of new objects	3-7 days	3 min-5 hr
Water requirements	daily	2-4 days without
Food per day	28 gm (1 oz)	3 gm (0.1 oz)
Water	57 gm (2 oz)	3 gm (0.1 oz)
Favourite foods	rolled oats, meat, fish, vegetable oil	grains, rolled oats, sugar, raisins
Droppings	bean size	rice size
Minimum width for entrance (hole diameter)	12 mm (0.5 in.)	6 mm (0.2 in.)
Can chew through (given edge to gnaw on)	rubber, alumunimum, cinder blocks, plastic, wool	same as rats

Rodent Control (The Principles)

Rodent control requires an integrated pest-management strategy involving many techniques. The producer's first objective should be to prevent, or at least greatly reduce, rodent numbers through management programs that eliminate entrance to the facility, nesting sites for the rodents, food supplies and water. Populations build when food, water and nesting sites are readily available.

Habits and Biology

To control mice and rats, we have to understand their habits and biology first. Mice and rats are similar in their habits and biology, although there are some differences between the two.

- Both are highly reproductive and extremely capable of surviving in all kinds of conditions. Theoretically, if there is one pair of mice (1 male and 1 female) in your barn at the beginning of a year, under fair living conditions, by the end of the year, you may have thousands of them on your farm.
- On farms, mice and rats will be near a food source such as barns, granaries, livestock buildings and silos.
- Rats and mice can climb and jump. Rats can jump vertically as high as 91 cm (36 in.) and horizontally as far as 122 cm (48 in.).
- Mice and rats can climb brick and other rough walls, and travel along utility wires.
- Rats can cross (sneak in) through openings as small as 1 cm (½ in.) and mice can squeeze through openings of 0.6 cm (¼ in.), or less, in diameter.
- Both mice and rats are active at night, particularly right after dusk.
- Rats are smart and tend to avoid new objects. Therefore, it may take a few days for traps and baits to work.

Rodent-Proofing Farm Buildings

Proper construction and maintenance of buildings helps prevent rodents from entering your barn. Initial construction footings should extend 0.5 m (19 in.) into the ground with an apron that extends 0.2 m (8 in.) outward. This prevents rodents from burrowing into your building. To prevent frost damage, footings may have to be deeper. Examine your building at least once a year for possible entry ways for rodents. Remember, a mouse needs only a 0.6-cm (¼-in.) opening to gain access; rats need a 1-cm (½-in.) opening. Cracks around door frames, under doors, broken windows, water and utility hook-ups, vents and holes surrounding feed augers are potential points of entry. Use coarse steel wool, hardware cloth or sheet metal to cover any entrances. Do not use plastic, wood or insulation, as rodents simply gnaw their way through.

When constructing walls, ensure that sheeting lies flush to the wall studs rather than on strapping. This keeps nesting sites confined to a single section between studs rather than allowing complete access to all wall spaces. A well-maintained structure is your first defence against rodents. Most rodents enter your barn directly from the fields, then the population builds. It is important to maintain good sanitation outside the barn. Eliminate vegetation for 1 m (3 ft) around buildings, clean up spilled feed, remove loose wood, garbage, etc. Do not attract rodents from fields to your operation.

Eliminating Hiding Places and Nesting Sites

Rodents do not like to be exposed. Maintain sound housekeeping, eliminate loosely piled building materials, old feed bags or anything else that a rodent can hide in or under. Keep piles of lumber, miscellaneous equipment 24-30 cm (9-12 in.) off the floor and at least 24 cm (9 in.) out from a wall. Look for entrances into double wall construction. Most rodents nest in the insulation of double walls. Block off all entrances into walls and destroy all nesting material.

Remove Food and Water

Eliminate water sources such as leaky taps, open water troughs, sweating pipes and open drains. Keep all feeds in rodent-proof bins, covered cans or metal hoppers. Reduce feed spillage and immediately dispose of dead animals. Without readily available food and water, populations cannot build.

Control of Existing Population

If there is already a rodent problem inside the barns, prevention alone won't solve the problem. In this case, consider a population-reduction program.

Snap Traps

For small populations, snap traps or box traps are very useful for eliminating rodents. Rats prefer fresh bacon, fish and meat, while mice favour cheese, peanut butter or seeds. Try several baits to find out which your rodents prefer. Rats are distrustful of anything new in their environment, so leave baited non-set traps out for 4-5 days to allow them to get used to the traps.

Ensure that previous baits have been taken before actually setting the traps. If rats are the problem, use rat traps. If mice are the problem, use a mouse trap. Locate traps close to walls, behind objects, in dark corners, where you see droppings or gnaw marks. When trapping next to a wall, set the trap at right angles to the wall with the trigger and bait closest to the wall. Orient multiple-catch traps with the entrance hole parallel to the wall. Live traps can work very well near runways used by mice and rats.

Glue Boards

Glue boards are very effective against mice and are the method of choice in locations where toxic baits are a concern. Glue boards will not work well if there is too much dust. They are only recommended where dust can be kept away from them. Check glue boards and traps daily and remove and dispose of dead mice and rats. Wear rubber gloves when handling them to prevent any chance of disease infection.

Abundant food supplies make baited traps less effective. Eliminate as many sources of food as possible before starting the program. For barns and poultry houses with moderate infestations, set 50-100 traps. The trapping program should be short and decisive to prevent trap shyness. Odours from humans or previously caught rodents do not cause trap shyness. When disposing of dead rodents, use plastic gloves and place the rodents in tightly sealed plastic bags. The Ontario Ministry of Environment is responsible for regulating pesticide sale, use, transportation, storage and disposal in Ontario. Ontario regulates pesticides by placing appropriate education, licensing, and/or permit requirements on their use, under the *Pesticides Act* and Regulation 63/09.

All pesticides must be used in accordance with requirements under the *Pesticides Act* and Regulation 63/09. Ontario farmers, as defined by the *Pesticides Act* and Regulation 63/09, may purchase and use Class 4, 5, and 6 pesticides for rodent control. Under Regulation 63/09 of the *Pesticides Act*, Ontario farmers must be certified to purchase and use Class 3 pesticides for rodent control on their farms. To be certified, a farmer must successfully complete the Grower Pesticide Safety Course. Certified farmers may also purchase Class 4, 5, and 6 products. Class 5 and 6 pesticides for rodent control can be used by homeowners.

Table: *List of approved active ingredients for rodent control in Ontario*, May 2010*

Active Ingredient	*Ontario Approved Class(es)*
brodifacoum	4,6
bromadiolone	4,5,6
bromethalin	3
cellulose from powdered corn cobs	4,5,6
chlorophacinone	4,5,6
difethialone	4,6
diphacinone	4,5,6
warfarin and sulfaquinoxaline	4,6
warfarin	3,4,5,6
zinc phosphide	3

* excludes Class 1 products used by manufacturers

There are two basic types of rodenticides: acute poisons and anti-coagulants. Use rodenticides when control of moderate-to-large rodent populations is necessary. Many of the newer anti-coagulant products, i.e., bromadiolone and brodifacoum require single feedings by rodents to cause mortality. Occasionally, rodents may develop a bait shyness after being made sick but not killed by a rodenticide. The shyness develops to the bait carrier, e.g., grain, and not to the rodenticide. Simply use another formulated product or different attractant if bait shyness develops. For rats, prebait using baits without the poison for about 1 week to get them accustomed to the bait. Place baits in areas of high rodent activity. Many people underbait in their control program. Baits should be 1-2 m (3-6 ft) apart for mice and 7-10 m (23-33 ft) for rats. Remove all uneaten baits and properly dispose of them after the poisoning program.

Precautions When Using Rodenticides

Ideally, cover all baits to prevent consumption by children, cats, dogs, poultry and swine. This can be done by placing baits in bait stations or bait boxes that allow ready access by rodents but prevent larger animals from gaining access. A baiting station designed from PVC piping has proven very effective in reducing rodent numbers. The cap and interior baffles keep the bait from falling out.

The advantages of this station include:

- station is easy to fill and replenish
- bait stays dry, making baits more attractive to rodents
- small entrance attracts rodents and increases feeding, particularly if plenty of additional feed is available
- design prevents feeding by pets, birds and livestock

Ensure that pets, livesto[ck] and poultry have no access to water *bait stations*.

Groundhog

The groundhog (Marmota monax), also known as a woodchuck, whistle-pig, or land-beaver in some areas, is a rodent of the family Sciuridae, belonging to the group of large ground squirrels known as marmots. Other marmots, such as the yellow-bellied and hoary marmots, live in rocky and mountainous areas, but the groundhog is a lowland creature. It is widely distributed in North America and common in the northeastern and central United States as well as Canada. Groundhogs are found as far north as Alaska, with their habitat extending southeast to Georgia.

Description

The groundhog is the largest sciurid in its geographical range, typically measuring 40 to 65 cm (16 to 26 in) long (including a 15 cm (6 in) tail) and weighing 2 to 4 kg (4 to 9 lb). In areas with fewer natural predators and large amounts of alfalfa, groundhogs can grow to 80 cm (30 in) and 14 kg (31 lb). Groundhogs are well adapted for digging, with short but powerful limbs and curved, thick claws. Unlike other sciurids, the groundhog's spine is curved, more like that of a mole, and the tail is comparably shorter as well—only about one-fourth of body length. Suited to their temperate habitat, groundhogs are covered with two coats of fur: a dense grey undercoat and a longer coat of banded guard hairs that gives the groundhog its distinctive "frosted" appearance.

In the wild, groundhogs can live up to six years, with two or three being average. In captivity, groundhogs reportedly live from 9–14 years. Common predators for groundhogs include wolves, cougars, coyotes, foxes, bobcats, bears, eagles, and dogs. Young groundhogs are often at risk for predation by snakes, which easily enter the burrow.

Figure: *Groundhogs are able to climb trees to escape*

The etymology of the name woodchuck is unrelated to wood or chucking. It stems from an Algonquian (possibly Narragansett) name for the animal, wuchak. The similarity between the words has led to the popular tongue-twister:

How much wood would a woodchuck chuck

if a woodchuck could chuck wood?

A woodchuck would chuck all the wood he could
if a woodchuck could chuck wood!

Diet

Mostly herbivorous, groundhogs primarily eat wild grasses and other vegetation, including berries and agricultural crops, when available Groundhogs also eat grubs, grasshoppers, insects, snails and other small animals, but are not as omnivorous as many other Sciuridae. Like squirrels, they also have been observed sitting up eating nuts such as shagbark hickory, but unlike squirrels, do not bury them for future use. Groundhogs hydrate through eating leafy plants rather than drinking from a water source.

Behaviour

Groundhogs are one of the few species that enter into true hibernation, and often build a separate "winter burrow" for this purpose. This burrow is usually in a wooded or brushy area and is dug below the frost line and remains at a stable temperature well above freezing during the winter months. In most areas, groundhogs hibernate from October to March or April, but in more temperate areas, they may hibernate as little as three months. To survive the winter, they are at their maximum weight shortly before entering hibernation. They emerge from hibernation with some remaining body fat to live on until the warmer spring weather produces abundant plant materials for food. Groundhogs are mostly diurnal, and are often active early in the morning or late afternoon.

Despite their heavy-bodied appearance, groundhogs are accomplished swimmers and excellent tree climbers when escaping predators or when they want to survey their surroundings. They prefer to retreat to their burrows when threatened; if the burrow is invaded, the groundhog tenaciously defends itself with its two large incisors and front claws. Groundhogs are generally agonistic and territorial among their own species, and may skirmish to establish dominance.

Outside their burrow, individuals are alert when not actively feeding. It is common to see one or more nearly-motionless individuals standing erect on their hind feet watching for danger. When alarmed, they use a high-pitched whistle to warn the rest of the colony, hence the name "whistle-pig". Groundhogs may squeal when fighting, seriously injured, or caught by a predator. Other sounds groundhogs may make are low barks and a sound produced by grinding their teeth. When groundhogs are frightened, the hairs of the tail stand straight up, giving the tail the appearance of a hair brush.

Small Mammals

Chipmunks & Squirrels

Chipmunks are small, striped rodents of the family Sciuridae. All species of chipmunks are found in North America, with the exception of the Siberian chipmunk, which is found primarily in Asia.

These furry rodents are regulars at many backyard birdfeeders and will also attack a variety of garden targets such as young seedlings, berries, fruits, and vegetables. They even have been known to decapitate flowers such as tulips — seemingly just for fun. Chipmunks are particularly good at unearthing bulbs, eating freshly planted seeds, and munching on maturing root crops.

Diet

Figure: *An eastern chipmunk eating*

Chipmunks have an omnivorous diet primarily consisting of seeds, nuts and other fruits, and buds. They also commonly eat grass, shoots, and many other forms of plant matter, as well as fungi, insects and other arthropods, small frogs, worms, and bird eggs. Around humans, chipmunks can eat cultivated grains and vegetables, and other plants from farms and gardens, so they are sometimes considered pests. Chipmunks mostly forage on the ground, but they climb trees to obtain nuts such as hazelnuts and acorns. At the beginning of autumn, many species of chipmunk begin to stockpile nonperishable foods for winter. They mostly cache their foods in a larder in their burrows and remain in their nests until spring, unlike some other species, which make multiple small caches of food. Cheek pouches allow chipmunks to carry multiple food items to their burrows for either storage or consumption.

Control

These rodents are smart. Given time, they'll find a way to get to their desired target. As with birds, it's best to vary your approach every so often to keep the rodents off balance. An good exclusion trick is to place a baffle on fruit tree trunks or bird feeder poles. New 3 squirrel-proof3 feeders appear frequently in catalogs — they may feature counter-weighted or electrified perches that surprise the critters when they visit the feeder. Some gardeners have had success spraying animal repellents such as cayenne pepper mix, predator urine, mothballs, and commercial formulas on or around plants. Others have had short-term success using ultrasonic noisemakers or setting up separate squirrel-feeding stations to lure them away from the garden. Trapping is the surest way to remove problem rodents. Live traps are very effective for particularly pesky squirrels or chipmunks. However, don't expect to control a whole population with trapping — there are always more squirrels and chipmunks waiting to move in.

Mice, Voles, & Rats

From the smallest house mouse to the largest Norway rat, these rodents live wherever humans do. Rats and mice are active at night, while voles can be seen scurrying about day or night. These creatures love to dig and burrow, making them major pests of underground plantings in the garden. Partially eaten potatoes or carrots are a sign of mice and vole activity. They also devour newly sown seeds, and feed on most flower bulbs (except daffodils). Spring snowmelt often reveals bark chewed from the base of young tree trunks.

Control

The first line of defence is to create a mouse-unfriendly environment by removing food, water, and shelter. Seal pathways they might use to enter buildings, clean up pet food dishes, and keep garbage in secured bins. The most effective household control is the old-fashioned trap. Whether it be a snap trap or live trap, bait it with peanut butter or bacon and place it in mouse runways. In the garden, delay using fluffy mulches such as straw until late in the season. Mice love to winter over in this material and it gives them easy access to your plants under the snow. Repellents containing ammonium can be effective against rodents when applied as directed. Protect tree trunks with tree wrap or hardware cloth, making sure it reaches a few inches below the soil line and 18 inches high. Place precious bulbs such as lilies in metal cages or place sharp, crushed gravel around the bulb when planting to deter these tunneling pests.

Opossums

Opossums make up the largest order of marsupials in the Western Hemisphere, including 103 or more species in 19 genera. They are also commonly called *possums*, though that term technically refers to Australian fauna of the suborder Phalangeriformes. The Virginia opossum was the first animal to be named an *opossum*; usage of the name was published in 1610. The word *opossum* was borrowed from the Virginia Algonquian (Powhatan) language in the form *aposoum* and ultimately derives from the Proto-Algonquian word **waÿp- a"èemw*, meaning "white dog" or "white beast/animal".

Their unspecialised biology, flexible diet, and reproductive strategy make them successful colonizers and survivors in diverse locations and conditions.

Let's face it, opossums aren't pretty. They have up to 50 needle-like teeth and ooze offencive fluids from both ends of their body. But like most of the creatures we've profiled here, they learn fast. These nocturnal marsupials are common east of the Rockies and along the West coast. They like forested areas with streams or swamps nearby, although like raccoons, they are increasingly found in urban areas. Opossums' favoured food is road kill (which they often become because of their nighttime activity), but they'll also partake of fruits and vegetables, especially tomatoes and corn. And like raccoons, opossums rummage through trash cans. If you see a mauled tomato or corn plant with fruit missing, opossums may be your culprit.

Control

To discourage opossums from visiting your yard, stow water and food supplies, such as pet bowls and bird feeders, out of reach at night.

Opossums also can be deterred with repellents, scare tactics, and a good fence. Moderately successful homemade repellents include a dusting of baby powder, cayenne pepper, or blood meal on vegetables and corn ears. In a small garden, cover individual corn ears or melons with bags taped shut. Place baffles on the trunks of fruit trees to prevent opossums from climbing them. Prune overhanging limbs that opossums might use to launch themselves into the fruit tree. To prevent lawn damage, control grubs and other insects. You can scare opossums temporarily with loud, intermittent noises or lights, but the racket generated by motion-sensor noisemakers or lights may prevent you from getting much sleep at night! The best deterrent is a good fence. Opossums are agile and smart, so build your fence well. If all else fails, remove persistent marauders with a trap.

Raccoons

The raccoon, sometimes spelled racoon, also known as the common raccoon, North American raccoon, northern raccoon and colloquially as coon, is a medium-sized mammal native to North America. The raccoon is the largest of the procyonid family, having a body length of 40 to 70 cm (16 to 28 in) and a body weight of 3.5 to 9 kg (8 to 20 lb). Its grayish coat mostly consists of dense underfur which insulates against cold weather. Two of the raccoon's most distinctive features are its extremely dexterous front paws and its facial mask, which are themes in the mythology of several Native American tribes. Raccoons are noted for their intelligence, with studies showing that they are able to remember the solution to tasks for up to three years. The diet of the omnivorous raccoon, which is usually nocturnal, consists of about 40% invertebrates, 33% plant foods, and 27% vertebrates.

Figure: *The mask of a raccoon is often interrupted by a brown-black streak that extends from forehead to nose.*

The original habitats of the raccoon are deciduous and mixed forests, but due to their adaptability they have extended their range to mountainous areas, coastal marshes, and urban areas, where some homeowners consider them to be pests. As a result of escapes and deliberate introductions in the mid-20th century, raccoons are now also distributed across the European mainland, the Caucasus region and Japan.

Though previously thought to be solitary, there is now evidence that raccoons engage in gender-specific social behaviour.

Raccoons are nocturnal animals that are thriving thanks to increased human contact, as anyone with an unsecured dumpster or trash can will attest. Wild populations prefer areas with trees and water nearby. But more and more they are found roaming city streets at night due to man's generosity in providing new sources of food, water, and shelter. Beside garbage cans, raccoons are notorious for their love of sweet corn. These omnivores will also eat strawberries, tree fruits, peas, potatoes, melons, and grubs found in your lawn. They will dig holes in maturing melons and munch on ripening tree fruit. They make holes in lawns and mulch piles looking for insects to eat. Like squirrels, they also have been known to clean out bird feeders.

Control

To discourage raccoons from visiting your yard, stow water and food supplies, such as pet bowls and bird feeders, out of reach at night. You can also deter raccoons with repellents, scare tactics, and a good fence. Successful homemade repellents include a dusting of baby powder, cayenne pepper, or blood meal on vegetables and corn ears. In a small garden, cover individual corn ears or melons with bags taped shut. Place baffles on the trunks of fruit trees to prevent raccoons from climbing them. Prune overhanging limbs that raccoons might use to launch themselves into the fruit tree. To prevent lawn damage, control grubs and other insects. You can scare raccoons temporarily with loud, intermittent noises or lights, but the racket generated by motion-sensor noisemakers or lights may prevent you from getting much sleep at night! The best deterrent is a good fence. Raccoons are agile and smart, so build your fence well. If all else fails, remove persistent marauders with a trap.

Skunks

These nocturnal critters are found throughout the country, especially near water. Their diet consists mostly of small animals and insects. They are mild-mannered creatures and will only shoot their notorious spray when threatened. Skunks only occasionally bother the garden, perhaps feeding on corn that's close to the ground. Mostly, their damage is due to night-time digging in lawns and flower beds as they look for one of their favourite foods: grubs. Skunks may dig many 3- to 4-inch-deep holes and make a lawn look like a war-zone in one night.

Control

The two best ways to rid yourself of skunks are to eliminate their food source or trap them. Spray beneficial nematodes or milky spore disease on lawns and flower beds in spring to reduce the grub population. You can also place ammonia-soaked rags in the skunk's den area (usually in an old woodchuck hole or under an outbuilding) to ward it off. If you have a problem skunk under a porch, woodpile, or stone wall, it's best to trap and relocate it.

Armadillo

Getting rid of armadillos is no longer a problem reserved for Texans. They were first seen in the Lone Star State in the 1850s and over the next hundred years, they'd waddled their way to Alabama and beyond. Armadillo control has become a concern throughout the southwest and beyond. Eventually, they'll be found in any state where winters are mild. They're known for tearing up flower beds in search of bugs and worms and leaving 3×5 inch divots in the lawn where they've dug up the turf looking for grubs. Before you ask about how to get rid of armadillos, you need to know a little about them.

The nine-banded armadillo (*Dasypus novemcintus*) is nocturnal, which means it does most of its foraging at night. Its strong legs and claws are built for tearing apart termite mounds and digging burrows that can reach 15 feet long. They eat bugs and grubs and worms, but the claim that they carry and spread leprosy is largely unprovable and unfounded. One of the reasons getting rid of armadillos is so difficult is that they aren't territorial. The one that's in your yard today may not be the one that did all that damage last week.

How to Stop Armadillos in the Garden

Unfortunately, the best method to stop armadillos from entering your yard is not only the most expensive, but might also be the least attractive. A stout fence with no spaces big enough for the critters to crawl through and buried a foot or more underground so they can't dig under it, is the best form of armadillo control.

But if you're not agreeable to living inside a fenced fortress, using their own biology against them might be a more practical and effective method of getting rid of armadillos. Armadillos have a great sense of smell and a large part of their brain is dedicated to it, so the answer to how to get rid of armadillos is fairly simple. Make your yard stink! Yes, strong scented, eye-stinging scents like those of vinegar or ammonia or good old pine cleaner can stop armadillos in their tracks, driving them from their borrows and your yard. Rumor has it, these roly-poly creatures are offended by the smell of pine needles or pine bark. You might try switching to one of these as mulch for your garden beds.

There is no repellent currently registered for armadillo control although there are several ultrasonic pest devices that claim to do much the same thing.

Trapping and Killing Armadillos

If easier, less confrontational methods fail, you might want to try trapping your midnight visitors. There are several devices available that are designed to capture without killing. Armadillos are partial to over ripe fruit and earthworms as bait. Try setting out a dish of bait for several nights before loading the trap to capture their interest first.

Killing armadillos may be your last and only solution to ridding your yard of this nocturnal pest. These animals are so focused on their search for food they notice little else, including flashlights and people! If you choose this method of getting rid of armadillos, make sure you check local ordinances governing the use of firearms and weapons.

As you can see, there are a variety of methods to stop armadillos from destroying your yard. Test them all and see which works best for you.

Large Mammals

Coyotes

Once confined to just the plains, this solitary canine is now found throughout the United States. Coyotes have learned to live with man by being active when man is not—when it's dark—and their presence may be more of a concern to pet and livestock owners. Since coyotes diet consists mostly of small animals and rodents, they could help your garden more than hurt it, but they can do their share of damage in the garden as well. Coyotes bear young in spring. At that time, they are particularly active and aggressive while looking for food and protecting their young. They aren't picky eaters. In addition to small animals, they eat fruits such as pears, and vegetables such as melons. Look for tooth marks and dog-like tracks to determine if Wile E. Coyote has been visiting your garden.

Controls: If you're concerned about coyotes in the neighbourhood, cut down weedy areas and tall grass areas where their main food (small rodents) live and where coyotes can hide. Secure garbage can lids and remove pet dishes at night. Cover vegetables at night and pick up windfall fruits, or use fencing to exclude coyotes from the garden. Coyotes are not easy to scare off, but may be frightened by large guard dogs, bright lights, or sounds.

Getting Rid of Foxes in the Garden

Many of us are familiar with wildlife pilfering our gardens' bounty, usually any number of birds and deer are the culprits. In some areas of the country, however, the outlaw's name is...the fox. Let's learn more about how to prevent foxes in the garden.

While some people count foxes as rather endearing, cute even (that'd be me) fox pest control may be a serious issue in the garden. Foxes are often an introduced, non-native, species which can disturb the delicate balance of an ecosystem. Over time escapees introduced for the purposes of fox hunting and fur farming, roamed free and comfortably settled in coastal and valley ecosystems. Prey for the fox are rodents, rabbits, reptiles, bird eggs, insects, waterfowl and other ground nesting birds, and they make no differentiation between imperiled species. There are several types of fox found in North America: the swift fox, kit fox, Arctic fox, gray fox and red fox – with the latter usually being the trouble maker. The red fox is the most widely distributed carnivore in the world, adapting easily to a variety of habitats.

Why Prevent Foxes in the Garden

Keeping foxes away from gardens may be important for safety and fiscal reasons. Although the fox is a solitary animal and usually eats small mammals and birds, piglets, kids, lambs and poultry ranging and foraging amongst your garden are just as enticing, especially when this may seem to be a fairly easy meal for these opportunists. Replacing the hen house occupants over time can be costly.

Rabies, although on the decrease, is also a concern and can potentially affect humans, domestic livestock and wildlife. Not forgetting, of course, the effect a fox in the garden will have on the songbirds you awaken to. So, our question stands, "how to deter foxes from gardens?"

Getting Rid of Foxes in the Garden

Getting rid of foxes in your garden can be accomplished by the simplicity of fencing. A net wire fence with openings of 3 inches or less and buried to a depth of 1 or 2 feet with an apron of net wire extending one foot outward from the bottom is a definite fox deterrent. You may take it a step further and include a roof of net wire as well. Additionally, an electric fence, spaced 6, 12, and 18 inches above ground will also repel foxes or a combination of both the net wire and electric fence. With repetition, foxes adapt to loud noises, however temporarily. Noise making devices can deter the fox activity as will flashing lights (strobe lights). In conjunction at irregular intervals, they are satisfactorily effective in the short term. The barking of the family dog will also be of some assistance in getting rid of foxes. Lastly, if you can really make no headway in ridding the garden of foxes, call in an expert who can safely trap and remove the animal.

Additional Fox Pest Control

Foxes in the small home garden are really a nuisance and the above solutions will probably solve the issue. There are other more deadly options that are not necessarily recommended for a home gardener. They are normally utilised by commercial producers of livestock and poultry, whose livelihood is directly affected by fox predation. These methods include shooting, fumigation with gas cartridges, poisoning via sodium cyanide, trapping, and den hunting. Most states allow the taking of foxes to protect private property but check with your state wildlife agency for regulations.

Deer

Although most people think of Bambi as a cute forest creature with retiring behaviour, due to an growing population, deer have become a major garden pest throughout the country. Although they tend to keep to forest edges and fields grazing on grasses and leaves, they become more daring when food is scarce, venturing into suburban yards. Deer graze and browse leaves, stems, and buds of many woody plants, as well as alfalfa, roses, corn, vegetables, and fruits. Their damage is evident because they leave jagged leaf edges on the eaten plants, not to mention distinctive cloven hoof prints and bean-shaped droppings.

Control: Although deer will eat anything if hungry enough, given a choice they tend to stay away from succulent plants, poisonous plants, pungent flavoured plants, and plants with hairy or furry leaves. Plant ornamentals with these qualities in areas of heavy deer traffic. Some gardeners have had success using human hair, dog hair, soap, blood meal, rotten eggs, hot pepper, or predator urine spread around or on flowers and trees. Deer can be scared away by motion sensor devices attached to lights or loud music. Of all the methods, though, fencing is the most reliable. It's best to erect the fence before Bambi has found your garden or yard.

Others

Birds

Many birds are friends to the gardener, as they feast on harmful insects and weed seeds. But when they feast on our crops, we categorize them as pests. Blackbirds, starlings, and even robins have been known to pull up tasty corn seedlings, munch on ripening cherries, peaches, and blueberries, and even dig holes in lawns looking for insects. The amount of damage will vary depending on the weather and native food supply that year. In dry years birds will attack many different fruits and vegetables, such as tomatoes and melons, for both food and moisture. For the birds, the timing couldn't be better. As gardeners are planting their crops and nurturing them to ripeness, these birds are raising their young, and need a handy food source. Flocks of birds can quickly decimate an entire berry patch or grain field, while individual birds can be very aggressive and determined to attack a particular crop.

Control: Any scare device will work for a few days, but birds are smart and will soon get used to it. Some effective visual scare methods include hanging aluminum pie tins, reflective Mylar tape, or large scare eye balloons (that mimic the scary eyes of predator birds). Noisemakers, such as radios or intermittent canon exploders, are also useful. Commercial bird repellents and homemade hot pepper spray applied to favoured fruits are effective, too. However, the best way to control bird damage is through exclusion. Bird netting protects tree fruit and berries if installed 2 to 3 weeks before fruit mature. You can also cover maturing corn ears with paper bags, and lay floating row covers over germinating seedlings to foil the birds.

Cats & Dogs

Man's best friends can also be your garden's worst enemy. An untrained dog can maul plants or dig up freshly planted bulbs, flowers, or your lawn. Cats won't do as much digging and damage in general, but they love to lie on freshly turned earth where your vegetable seeds were just planted. Both animals can use a garden and lawn as a litter box, making for dead spots in the lawn and very unpleasant surprises for you in the garden. Another concern is that male dogs need to "mark their turf." If your lettuce or greens have a strange ammonium smell, think twice about eating them. Female dogs are usually responsible for brown spots on the lawn.

Control: Obviously the best control is to give pets their own yard or train them well. You can also repel cats and dogs with commercial sprays or homemade recipes. A solution of garlic, hot pepper, soapy water, or commercial repellent sprayed on prized plants may teach cats and plodding dogs to go elsewhere. Cats hate the smell of rue and citronella, so sprinkle these herbs in the garden. Beware, though — some repellents may encourage a dog to urinate on the spot more often to mark over the strange smell. A patch of catnip planted at a distance may lure felines away from your yard. In a new seedbed, lay down bramble stems or wire mesh to prevent cats from using it as a litter box. For dogs, fence an area that can be used as a bathroom, and be diligent about watering areas where dogs urinate regularly to flush salts from the soil. For sure-fire protection, erect a wire fence with sturdy stakes.

Snakes

Although popular lore suggests snakes should be feared or exterminated, relatively few are poisonous and they generally avoid humans. Besides, they can be the gardener's friend, as various snakes prey upon rodents and slugs. Seeing snakes in your yard is a sign that you have a healthy population of mice and voles. To encourage snakes to inhabit your yard and garden, provide them with a shallow dish of water and many cool spots to hide. Rock walls, woodpiles, or boards laid in the garden provide shelter and a place to soak up the sun's warming rays.

Control: Even though snakes are 3 good3 guys in the garden, no one likes them in the house, so patch any cracks into your house or basement. Snakes particularly like the relief of a cool cement basement on a steamy summer day. If one does find its way indoors, call a wildlife control officer to remove it. If your fear of snakes is overwhelming, try discouraging them by removing weeds, woodpiles, rock walls, water sources, and brush where mice might hide. Commercial repellents or a fence can also be used to exclude them from a favoured area.

Some Other Control Measures

Animal Fencing

Fencing is the only sure-fired way to keep wild and domestic animals out of a prized garden long term. Although it can be expensive for large animals such as deer, it may be the only way to protect your yard from costly damage. Here are some points to keep in mind when planning to build a wildlife fence.

- Check local ordinances on building fences. Large fences may block desired views.
- Polypropylene fencing is less obtrusive than solid fencing, but provides protection only from certain animals.

- Fences work best when the animal doesn't know what it's protecting. A hungry animal that knows food is available will try harder to get through any fence. Solid fences provide a visual block so they don't know what they're missing.
- Electric fences are more expensive and need maintenance, but they make less of an impact on your view and can be more effective than solid fences.
- Electric fences work best if animals know they are "hot". Smear peanut butter on the fence as soon as you put it up to entice animals to get shocked. They'll be less likely to test the fence again. Keep vegetation from touching the fence. It will reduce its charge or ground it out.

Deer

Deer are probably at the top of the list of animals home owners want to fence out. Since deer can jump, the fence needs to be constructed high and at an angle to deter them. There are many design variations for deer barriers available from game wardens. One effective option is an 8- to 10-foot-tall fence slanted at a 45 degree angle toward the direction from which deer are most likely to come. It will make them think twice about jumping. Keep the fence snug to the ground, since deer can also wiggle under fences. Electric fences baited with peanut butter and solid fences that block the view to a food source also work well.

Cats and Dogs

Build a wire mesh fence 3 feet high anchored with sturdy posts. Cats probably won't climb over, and most dogs can't knock it over. Bend the base of the fence outward to form a 2-foot-wide apron along the ground to discourage dogs from digging under it.

Rabbits

Exclude rabbits with a 2-foot-tall chicken wire fence that has 1-inch-diameter holes. To prevent them from digging under, curve the bottom of the fence 90 degrees to create an apron a foot or so wide, and bury it several inches deep.

Woodchucks

Like raccoons and opossums, woodchucks are good climbers, so leave the top 18 inches of fence unattached, or string electric wire across the top to discourage these pests. The 4-foot-tall fence should also have a 2-foot-wide apron buried a few inches below the soil to stop the pests from burrowing under the fence. Electric fencing placed a few inches outside a wire fence also helps.

Animal Trapping

If you've tried repelling, excluding, and scaring away a pest animal and it's still causing problems, you can trap and release it, or use a lethal trap. Trapping is an effective way to remove a specific pest animal, but isn't useful against a large local population over the long run. In the latter case, trapping should be done in conjunction with exclusion, fencing, or altering the habitat.

It's best to call your local wildlife officer or professional pest removal company to handle problem animals. However, if you're willing to do the dirty work yourself, here are some tips to follow.

- Minimize signs of human activity. Handle traps as little as possible and mask your scent by wearing gloves. Make sure the trap is the right size and sturdy enough to handle the animal you're after.
- Anchor the trap so the animal can't roll it over and escape.
- Bait the trap a few times before setting to teach the animal that it's a safe to approach it.
- Make a trail of goodies to the trap, and pile some in the back or on the trap for the critter to find.
- Disinfect the trap with bleach after using. Rabies is transmitted by saliva, and you don't necessarily have to be bitten to contract the disease.
- Place plywood under live traps to prevent animal from scratching up the lawn or garden once they're caught.
- Use caution when releasing large animals (e.g. raccoon, skunk, woodchuck). The safest method is to tie a strong rope to the trap door. Run the rope through an open window into your vehicle, so you can pull open the trap door from a point of safety.
- Check traps daily, and release or bury animals as soon as possible.

Live Trapping

Live trapping and releasing animals has become popular. It seems no one really wants to hurt those fuzzy, furry, cuddly creatures. But in some cases live trapping can be more cruel than killing the animal. Relocated animals frequently become disoriented and aren't able to fit into their new habitat, often starving to death. If you use live traps, contact the Humane Society or the local or state game officer to determine the best method, time to trap, and way to release a wild or nuisance animal before setting the trap. Some species are protected from trapping by law in various states.

Once you've caught the animal, be careful. Wild animals can carry diseases such as rabies. Frightened skunks will be more likely to spray, and an agitated woodchuck will nip in self-defence. Here are some tips for trapping specific animals.

Chipmunks and Squirrels

Bait the trap with nuts, peanut butter, rolled oats, or seeds. Set traps near a food source (e.g. a bird feeder) or along a runaway such as at the base of a tree. Use metal traps, as rodents have sharp teeth and can chew through wooden traps. Check traps a few times a day.

Crown Borer Management: Treatment and Control of Crown Borers

When your garden starts to lgok a little scraggly and plants begin to die, any good gardener will check them all over for clues to the perpetrator. When you find holes in the base of trunks or canes with sawdust-like material coming out, your problem is most likely crown borers. Lets find out more about crown borer damage and control.

What are Crown Borers?

You may not realise you're looking for crown borer anformataon when you search Google, tryifg to discover the identity of the creature drilling holes in your caneberries afd ornamental plants, but, as it turns out, this damage is their most distinctive sign. The larvae of these clearwing moths tunnel their way into plants, eating as they go.

The life cycle of crown borer moths begins when adults emerge in June and July to lay their eggs on wounded or stressed plants, either on the bark or on leaves nearby. The larvae hatch and make their way to the crown, forming an overwintering area at the base of the plant that may have a blistered appearance. In their first spring, crown borer larvae begin to tunnel into the plant's crown, feeding until winter approaches, then making for the root system. After overwinterang as a larvae, they return to the crown and feed ravenously. Near the beginning of the second summer, these larvae pupate for two to four weeks, then emerge as adults to start the cycle again.

Crown Borer Management

Crown borer damage can be very non-specific, causing plants to wilt or appear sickly. Often the sawdust-like grass is the only sign of what's going on inside the crown. Adults, which look similar to black and yellow wasps, can be seen for short periods$ but may not mace themselves obvious in the landscape.

Because of this, control of crown borers is primarily defencive – infested plants should be removed as soon as possible to prevent the borers from spreading further. Prevent borers in new plantings by destroying wild brambles and other infected plants in t`e area and replanting with certified pest-free nursery stock. Borers are often attracted to stressed plants, so proper care, water and pruning is vital to prevention. Familiarize yourself with the needs of each of your landscape plants and make sure to water them adequately as summer temperatures rise. Regular pruning and shaping to remove redundant branches and open the inside of the canopy is recommended.

Secrets for a High-Yield Vegetable Garden

Imagine harvesting nearly half a ton of tasty, beautiful, organically grown vegetables from a 15-by-20-foot plot, 100 pounds of tomatoes from just 100 square feet (a 4-by-25-foot bed), or 20 pounds of carrots from just 24 square feet. Yields like these are easier to achieve than you may think. The secret to superproductive gardening is taking the time *now* to plan strategies that will work for your garden. Here are seven `igh-yied strategies gleaned from gardeners who have learned to make t`e most of their garden space.

Build Up Your Soil.

Expert gardeners agree that building up the soil is the single most important factor in pumping up yields. A deep, organicaldy rich soil encourages the growth of healthy, extensive roots that are able to reach more nutrients and water. The result: extra-lush, extra-productive growth above ground.

The fastest way to get that deep layer of fertile soil is to make raised beds. Raised beds yield up to four times more than the said amount of space planted in rows. That's due not only to their loose, fertile soil but also to efficient spacing—by using less space for paths, you have more room to grow plants. Raised beds save you time, too. One researcher tracked the time it took to plant and maintain a 30-by-30-foot garden planted in beds, and found that he needed to spend just 27 hours in the garden from mid-May to mid-October. Yet he was able to harvest 1,900 pounds of fresh vegetables—that's a year's supply of food for three people from about 3 total days of work!

How do raised beds save so much time? Plants grow close enough together to shade out competing weeds, so you spend less time weeding. The close spacing also makes watering and harvesting more efficient.

Round Out Your Beds.

The shape of your beds can make a difference, too. Raised beds are more space-efficient if the tops are gently rounded to form an arc, rather than flat. A rounded bed that is 5 feet wide across its base, for instance, will give you a 6-foot-wide arc above it—creating a planting surface that's a foot wider than that of a flat bed. That foot might not seem like much, but multiply it by the length of your bed and you'll see that it can make a big difference in total planting area.

In a 20-foot-long bed, for example, rounding the top increases your total planting area from 100 to 120 square feet. That's a 20 percent gain in planting space in a bed that takes up the same amount of ground space! Lettuce, spinach, and other greens are perfect crops for planting on the edges of a rounded bed.

Space Smartly.

To get the maximum yields from each bed, pay attention to how you arrange your plants. Avoid planting in square patterns or rows. Instead, stagger the plants by planting in triangles (as shown here). By doing so, you can fit 10 to 14 percent more plants in each bed.

Just be careful not to space your plants *too* tightly. Some plants won't reach their full size—or yield—when crowded. For instance, when one researcher increased the spacing between romaine lettuces from 8 to 10 inches, the harvest weight per plant doubled. (Remember that *weight* yield per square foot is more important than the number of plants per square foot.) Overly tight spacing can also stress plants, making them more susceptible to diseases and insect attack.

Grow Up!

No matter how small your garden, you can grow more by going vertical. Grow space-hungry vining crops—such as tomatoes, pole beans, peas, squash, melons, cukes, and so on—straight up, supported by trellises, fences, cages, or stakes. Growing vegetables vertically also saves time. Harvest and maintenance go faster because you can see exactly where the fruits are. And upward-bound plants are less likely to be hit by fungal diseases thanks to the improved air circulation around the foliage.

Try growing vining crops on trellises along one side of raised beds, using sturdy end posts with nylon mesh netting or string in between to provide a climbing surface. Tie the growing vines to the trellis. But don't worry about securing heavy fruits—even squash and melons will develop thicker stems for support.

Mix It Up.

Interplanting compatible crops saves space, too. Consider the classic Native American combination, the "three sisters"—corn, beans, and squash. Sturdy cornstalks support the pole beans, while squash grows freely on the ground below, shading out competing weeds. This combination works because the crops are compatible. Other compatible combinations include tomatoes, basil, and onions; leaf lettuce and peas or brassicas; carrots, onions, and radishes; and beets and celery.

Succeed with Successions

Succession planting allows you to grow more than one crop in a given space over the course of a growing season. That way, many gardeners are able to harvest three or even four crops from a single area.

For instance, an early crop of leaf lettuce can be followed with a fast-maturing corn, and the corn followed by more greens or overwintered garlic—all within a single growing season.

To get the most from your succession plantings:

- Use transplants. A transplant is already a month or so old when you plant it, and so will mature that much faster than a direct-seeded plant (one grown from seeds sown in the garden).
- Choose fast-maturing varieties.
- Replenish the soil with a ¼-to-½-inch layer of compost (about 2 cubic feet per 100 square feet) each time you replant. Work it into the top few inches of soil.

Stretch Your Season.

Adding a few weeks to each end of the growing season can buy you enough time to grow yet another succession crop—say a planting of leaf lettuce, kale, or turnips—or to harvest more end-of-the-season tomatoes.

To get those extra weeks of production, you need to keep the air around your plants warm, even when the weather is cold, by using mulches, cloches, row covers, or coldframes.

Or give heat-loving crops (such as melons, peppers, and eggplants) an extra-early start by using two "blankets"—one to warm the air and one to warm the soil in early spring. About 6 to 8 weeks before the last frost date, preheat cold soil by covering it with either infrared-transmitting (IRT) mulch or black plastic, which will absorb heat. Then, cover the bed with a slitted, clear plastic tunnel. When the soil temperature reaches 65° to 70°F, set out plants and cover the black plastic mulch with straw to keep it from trapping too much heat. Remove the clear plastic tunnel when the air temperature warms and all danger of frost has passed. Install it again at the end of the season, when temperatures cool.

Organic Pest-Control Techniques

Once you've identified the troublemakers, you can control them with an assortment of organic pest-control products.

Floating Row Covers

This translucent, white, porous polyester fabric acts as an insect barrier, while letting in up to 80 percent of the available light. You can buy either lightweight or heavyweight types—you'll want to use the lighter one for controlling pests in summer, because it will keep out bugs without cooking your plants. The heavier reportedly traps more warmth and so is better for season extending.

The material is sold by the yard, generally in rolls 4 to 8 feet wide. You cut it to the length you need, then drape it over metal hoops, attach it to wooden supporting frames, wrap it around wire tomato cages, or simply lay it directly on your crops like a blanket. Important: You must secure the edges of the row cover with soil, U-shaped pins (either commercial or homemade ones crafted from wire coat hangers), boards, bricks, or rocks.

Use floating row covers as temporary barriers to get plants past critical stages, such as when they are seedlings or while the pest you are deterring is most active. Of course, you could keep the crop covered for its entire life span, although this isn't a good option for crops that require insect pollination.

Pests Controlled: Row covers are especially useful against mobile pests, including cabbage moths (imported cabbageworms), Colorado potato beetles, most aphids, Mexican bean beetles, flea beetles, squash bugs, and tomato hornworms. Combine row covers with crop rotation if you're dealing with pests that overwinter in the soil.

Pheromone Traps

Many insects produce powerful smells called pheromones that they use to lure the opposite sex. Scientists have duplicated several of these scents and used them to bait special traps for luring the target insect. But because these "sex" traps attract mostly male insects, they aren't very effective controls. They're useful as an early warning that a particular pest is moving into your area. When you find the first pests in your trap, you know it's time to launch your control strategies, such as putting your row covers in position and applying Bacillus thuringiensis.

Pests monitored: Pheromone lures are available for diamondback moths and moths that produce armyworms, cabbage loopers, corn earworms, European corn borers, tomato pinworms, and cutworms.

Sticky Traps

These traps—a rigid material of a particular colour that's coated with a sticky substance—are used to catch insects that are attracted to that colour. To be effective, the traps must be clean and sticky. Also, use at least one trap (hung at plant height and close to the plant) every 3 to 5 feet. You can buy packaged sticky traps or make them yourself. To make your own, use any rigid material of the right colour or that you can spray paint. Cut the material to size (4 ' 6-inch rectangles are the standard), and if needed, paint it the correct colour. Cover the trap with a plastic bag or clingy plastic wrap, then coat it with a sticky substance, such as Tangle-Trap. (The plastic wrap makes cleanup easy—when your trap is covered with bugs, just remove the plastic and rewrap the trap with a new piece of plastic. Then coat it with more sticky stuff.)

For cucumber beetles, use a mobile trap: Wrap clingy plastic food wrap around a white bucket or other large object, then coat the plastic with Tangle-Trap. Carry the bucket along the rows of vines, shaking and brushing the plants as you go. The beetles will fly up and stick to the traps. Pests controlled: Yellow traps attract whiteflies, fruit flies, male winged scales, leafhoppers, fungus gnats, midges, male winged mealybugs and leafminers, thrips, psyllids, and winged aphids. White traps lure whiteflies, plant bugs, cucumber beetles, and flea beetles. Light blue traps attract flower thrips, and red spheres attract the flies whose eggs hatch into apple maggots.

Insecticidal Soap

Insecticidal soap contains unsaturated long-chain fatty acids (derived from animal fats) that dissolve the cuticle (skin) of insects. Insecticidal soap sprays are commercially formulated products sold specifically for insect control.

To be effective, the insecticidal soap must come in contact with the insects while it's still liquid—it has no effect after it dries on the plants. Spray only on pests and try to avoid hitting beneficial insects with the spray. Caution: Insecticidal soap can burn some plant leaves. Test each type of plant before spraying the entire plant. Spray a few leaves, then wait 48 hours. If there's no damage, go ahead and spray the entire plant. Don't spray on hot days, and rinse the soap off your plants after a few hours if the plants are receiving a lot of sunshine. If you have hard water, mix the soap with distilled water to help the soap dissolve.

Pests controlled: Insecticidal soap sprays are highly effective against mites, aphids, whiteflies, and other soft-bodied insects as well as the softer nymph stages of some tough-bodied bugs.

Oil Sprays

Oil sprays work by suffocating pests. To be effective, the oil spray must hit the pest directly. Use "dormant" oils to kill insect pest eggs and disease spores on the bare branches of trees and shrubs during the dormant season. To treat growing plants, use a lighter-weight, more refined horticultural oil (called "summer," "supreme," or "superior" oil). Lighter oils evaporate more quickly than dormant oils and are less likely to damage plants.

To avoid plant damage, don't spray any plants suffering from moisture stress. Also, don't spray on very hot days. Test the spray on just a few leaves before you spray the entire plant. Wait 48 hours to make sure no leaf spotting or discolouration occurs. To minimize potential harm to beneficial insects, limit your spraying to small areas where you can see pests lurking, and leave a couple of unsprayed "refuges" for any good bugs you can't see. Protect nectar-feeding beneficials by not spraying during peak flowering times and by not spraying blossoms. Spray early in the morning, before bees become active. And if you plan to release beneficials, do it after you apply the oil spray.

Pests controlled: Use horticultural oils to combat aphids, mites, beetles, leaf miners, caterpillars, thrips, leafhoppers, and whiteflies.

Bacillus Thuringiensis

Bacillus thuringiensis (BT) is a naturally occurring bacterium found in the soil. There are many different types, and some can be used to kill a specific insect or class of insects. When a target insect takes a bite of a plant sprayed with the type of BT the insect is sensitive to, the insect gets infected and stops feeding. Inside the insect, the bacterium releases a protein that causes the pest to die within a few days.

Each type of BT is effective only on one specific insect (or group) and only on insects that actually eat it. However, that doesn't mean you can spray it indiscriminately. For example, the type that kills cabbage loopers can also kill the caterpillars of the beautiful butterflies you're trying to attract to your garden. Only spray it when you know you have a pest problem, and only spray the pest-infested plants.

Most formulations of this bacterium are sold as a liquid or wettable powder that you dilute with water and then spray on the plants you want to protect. Some products are sold in the form of dusts or granules that you dust directly on plants. Because BT usually is effective only against the nonadult stage of pest insects, you must time applications carefully. As soon as you spot the pest larvae, thoroughly coat the affected plants with the spray or dust. (For corn pests, deposit a little of the granular product into the whorl or on the corn silk.) Avoid spraying during the heat of day. BT breaks down a day or two after spraying, so you may need to reapply it if you're up against a severe infestation. As with all sprays or dusts, always wear goggles and a mask to prevent contact with the bacterium when you apply it to your plants (there have been a few reports of allergic reactions in those who have inhaled it). Pests controlled: The most common strain of the bacterium—BT var. kurstaki (sometimes called BT var. berliner)—kills hundreds of different kinds of caterpillars, including cabbage loopers, tomato hornworms, cabbageworms, corn earworms, European corn borers, and squash vine borers. BT var. tenebrionis (a new name—until recently this one was called BT var. san diego) kills Colorado potato beetles.

Parasitic Nematodes

Don't confuse these beneficial nematodes with destructive root-knot nematodes. Once inside a pest, parasitic nematodes release bacteria that kills the insect host within a day or two. Although these good nematodes occur naturally in the soil, there usually aren't enough of them in one place to control pests that have gotten out of hand in your garden. But you can buy them by the billions for use as a living—and organic, safe, and nontoxic—form of pest control.

The dormant nematodes are shipped in a moist medium, which you mix with water when you're ready to apply. When you receive a shipment, put the sealed container in your refrigerator until you are ready to use it (the nematodes will keep there for about 4 months). Try to use them as soon as possible, though; their effectiveness declines the longer you store them. Once the nematodes are mixed with water for application, they are only viable for a very short time. Use all of the mix within a few hours—don't try to save any of it. Apply the nematodes to moist soil that has reached a temperature of at least 60°F, either in the evening or when it's overcast, at a rate of about 23 million nematodes per 1,000 square feet. Thoroughly cover the area with the nematodes, then water them in. Exception: If your pest is in the plant (the squash vine borer or corn earworm), mix up a small batch of nematodes and use a garden syringe or eye-dropper to apply them just inside the tip of the ear of corn, or into the squash vine entrance holes.

Pests controlled: Nematodes attack and invade armyworms, corn earworms, squash vine borers, soil-dwelling grubs (including Japanese beetle larvae), weevils, root maggots, and cutworms (in their soil-dwelling stages).

Bibliography

Arthur W. Gilbert, Mortier F. Barrus and Daniel Dean: *Growing and Breeding of Potatoes*, Asiatic Pub, Delhi, 2006.

Ashraf, Syed Mahboob: *Handbook of Fruit and Vegetable Products*, Agrobios, Delhi, 2008.

Bahar A. Siddiqui and Samiullah Khan: *Plant Breeding Advances and in vitro Culture*, CBS, Delhi, 1997.

Bhutani, R.C.: *Fruit and Vegetable Preservation*, Biotech Books, Delhi, 2003.

Cronquist, A.: *The Evolution and Classification of Flowering Plants*, New York Botanical Garden, Bronx, New York., 1988

Crucefix D.: *Avocado Variety Selection for Export Development,* Roseau, CARDI, 1996.

Doijode S. D.: *Seed Germination in Fruits*, New Delhi, Malhotra Publishers, 1993.

Ferentinos L.: *Proceeding of the Sustainable Taro Culture for the Pacific Conference*, Honolulu, HITAHR, 1993.

Fortuner, R.: *Nematode Identification and Expert System Technology,* New York, Plenum Press, 1988.

Gilbert Wooding Robinson: *Soils : Their Origin Constitution and Classification*, Biotech Books, Delhi, 2005.

Gupta, S.K.: *Disease Problems in Vegetable Production*, Scientific, Delhi, 2006.

Hessayon D. G. Dr.: *The Vegetable Expert*, England, PBI, Publications, 1985.

James S Shoemaker: *Culture of Vegetable Growing*, Asiatic Pub, Delhi, 2006.

Jeffers P.: *Evaluation of Four Onion Varieties in Montserrat*, Plymouth, CARDI, 1992.

Kataria, T N: *Plant and Crop Physiology*, Pearl Books, Delhi, 2008.

Kumar, N.: *Breeding of Horticultural Crops : Principles and Practices*, New India Pub, Delhi, 2006.

Matsuoka H.: *Cultivation of Panicum Genetic Resources for Evaluation of Characteristics*, Tokyo, JICA, 1997.

Mellor, J. W.: *The New Economics Growth of Vegetable Production*, Ithaca, Cornell University Press, 1976.

Morris M. L. : *Maize Seed Industries in Developing Countries*, Colorado, Lynne Rienner, 1998.

Narang, R.K.: *Fruit and Vegetable Preservation Techniques*, APH Pub, Delhi, 2010.

Nobel, P. S.: *Physicochemical and Environmental Plant Physiology*, Academic Press, San Diego, 1999.

Oldham P.: *Cost of Production of Major Tree Crops in Dominica*, Roseau, Ministry of Agriculture, 1991.

P. Bhattacharyya and S.S. Purohit: *Organic Farming : Biocontrol and Biopesticide Technology*, Agrobios, Delhi, 2008.

Pramod P Mahulikar and Kshama M Chavan: *Botanicals as Ecofriendly Pesticides*, New India Pub Agency, Delhi, 2007.

Rana, M. K.: *Breeding and Protection of Vegetables*, New India Pub, Delhi, 2011.

Rehm Sigmund: *Multilingual Dictionary of Agronomic Plants*, Boston, Kluwer Academic Publishers, 1994.

Seshadri, V.S.: *Cucurbit Vegetables : Biology, Production and Utilization*, Studium Press, Delhi, 2009.

Singh, Kirti: *Emerging Scenario in Vegetable Research and Development*, Research Periodicals and Book Publishing House, Delhi, 2001.

Singh, Neeraj Pratap: *Basic Concepts of Vegetable Science*, International Book, Delhi, 2004.

Somasundaram Rajarathnam: *Advances in Preservation and Processing Technologies of Fruits and Vegetables*, New India Publishing Agency, Delhi, 2011.

Srivastava, A.P.: *Mechanization of Vegetable Production and Post-Harvest Management*, Westville Pub, Delhi, 2009.

Tripathi, Pramila: *Botanical Pesticides in the Management of Post-harvest Fruit Diseases*, Daya, Delhi, 2005.

Woolfe Jennifer A.: *The Potato in the Human Diet*, Cambridge, Cambridge University Press, 1989.

Yadav, D.S.: *Advances in Vegetable Production*, Researchco Book Centre, Delhi, 2005.

Index

H

I

K

L

M

N

O

P

R

S

T

V

W

❑❑❑